Commerce in Culture

Commerce in Culture: States and Markets in the World Film Trade

Andrew J. Flibbert

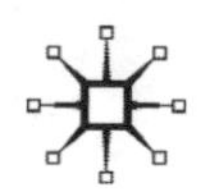

COMMERCE IN CULTURE

First published in 2007 by
PALGRAVE MACMILLAN™
175 Fifth Avenue, New York, N.Y. 10010 and
Houndmills, Basingstoke, Hampshire, England RG21 6XS.
Companies and representatives throughout the world.

PALGRAVE MACMILLAN is the global academic imprint of the Palgrave Macmillan division of St. Martin's Press, LLC and of Palgrave Macmillan Ltd. Macmillan® is a registered trademark in the United States, United Kingdom and other countries. Palgrave is a registered trademark in the European Union and other countries.

ISBN-13: 978-1-4039-8380-0
ISBN-10: 1-4039-8380-1

Library of Congress Cataloging-in-Publication Data

Flibbert, Andrew J.
Commerce in culture : states and markets in the world film trade / Andrew J. Flibbert.
p. cm.
Includes bibliographical references and index.
ISBN 1-4039-8380-1 (alk. paper)
1. Motion pictures—Economic aspects. 2. Motion pictures—Government policy. 3. Motion pictures—Marketing. I. Title.
PN1993.5.A1F54 2007
384'.83—dc22

2007003086

A catalogue record of the book is available from the British Library.

Design by Scribe Inc.

First edition: September 2007

10 9 8 7 6 5 4 3 2 1

Printed in the United States of America.

To Sonia and Alex

Contents

List of Figures and Tables

Figures

Tables

Preface

The inspiration for this book began on a London movie set in July 1995, when I took a daylong break from preparing for my PhD exams in political science to be an extra in an old college friend's short film. It was revealing to shift, even momentarily, from international relations theory to the profoundly creative yet technical domain of the film world. I was struck by how filmmaking is an extraordinary synthesis of art and industry. Filmmakers have no choice but to draw on and communicate with a socially shared repertoire of ideas, images, memories, and cultural knowledge, regardless of the extent of their artistic aspirations. They also operate by necessity, however, in a collaborative enterprise that cannot succeed without a broader industrial structure too easily taken for granted in the United States. This holds true even in the digital era, and for low-budget independent filmmakers and self-styled film *auteurs*, who never are as financially and institutionally unconstrained as artists and writers in other fields of creative endeavor.

Eventually, I concluded that the Janus-faced nature of filmmaking and its layers of interconnections are emblematic of broader transformations occurring in a globalizing world. These connections were especially significant to me as an experienced student of the Middle East, a region perennially regarded as *sui generis* and therefore subject to its own unique and inscrutable logic. Aware of Egypt's historical prominence as the dominant Arabic-language film producer, I soon discovered that Egypt has something in common with its Latin American counterpart, Mexico, which also has decades of film history. The many similarities between the two countries—and some crucial differences—presented an opportunity to explore a host of connections: over the space between Latin America and the Middle East; over time in the long sweep of film history; between the distinctive intellectual concerns of political economy, film studies, and area studies; and between the complementary subfields of comparative and international politics. Just as the study of political economy requires seeing, not obscuring, the interconnections between politics and economics, I decided to use filmmaking as a vehicle and venue for investigating connections, more than a unique form of cultural production itself.

This theme of connectedness extends to the book's core substantive claim that political and industrial structures shape state policies in the trade and cultural domains. State institutions matter in a relational sense—how they connect to each other—because they never exist and operate in isolation. State bureaucracies include both vertical hierarchies of authority and horizontal relationships between rival sets of decision makers. The latter fight turf wars, struggle for resources, identify interests, determine agendas, defend budgets, and make choices with one eye on other power centers in the state. Precisely how various bureaucratic actors relate to one another affects how state interests are defined and defended in a world of competing claims. Whether a state has a Ministry of Culture, for example, and the standing of that ministry in relation to its economic and financial counterparts can determine the precise criteria used to make policy decisions.

A similar story can be told about film industries, since their capacity to get what they want—protection from severe international competition, for example—hangs on the industrial structures organizing them. Market structures make and break industrial connections between specialized economic subsectors, as well as across national borders that wax and wane in significance. These structures can reward or preclude industrial cooperation, and they can mitigate or exacerbate fears of betrayal by segments of the industry that do not automatically share common interests. They are especially consequential in a globalizing world, where national industrial subsectors sometimes forge strategic alliances with international capital, marketing networks, and production chains. They affect the economic vitality of any given industry, as well as the nature and quality of the commodity it produces.

In filmmaking, this has been so for nearly a century. The industry's birth in the late nineteenth century led to an immediate global diffusion of its products and technology, as entrepreneurs sought to exploit the novelty of moving pictures for curious audiences everywhere. Whole systems of film production, distribution, exhibition, and even viewership spread throughout the world in short order, driven largely by the successes of the leading American companies and some of their European rivals. These systems connected to local industries and inserted them into the global capitalist economy. A few quick decades later, the advent of sound recording in film introduced the nationally specific element of language, which reinforced film's implication in the state- and nation-building projects ongoing worldwide. At its core, a political logic asserted itself early on and trumped the cultural and economic dynamics more conventionally identified with the seventh art. In a globalizing world, this political logic has remained central throughout the years, with little indication of its fading significance. How this can be so is the story told in the pages that follow.

Acknowledgments

Just as filmmaking is an act of creative and technical collaboration, so too is scholarly writing, and many people in academic and film circles have contributed to this book's completion in ways large and small. For years of unwavering intellectual support and guidance, I thank Lisa Anderson, Helen Milner, and Robert Vitalis, who each played essential and distinctive roles. Along the way, I also benefited from the insights, commentary, and introductions offered by Walter Armbrust, Douglas Chalmers, Harvey Feigenbaum, Greg Gause, Arthur Goldschmidt, Stephan Haggard, Steve Heydemann, the late Amy J. Johnson, Robert Kaufman, Mark Kesselman, Robert Kolker, Joseph Massad, Richard Peña, Jack Snyder, John Tian, John Trumpbour, and David Waldner. I am grateful to Jehan Sadat and her representative, Linda Woolridge, for Mrs. Sadat's recollections of July 1952. Columbia University provided exceptional support in material and intellectual terms.

I was fortunate as well to have had a Fulbright / Institute for International Education grant for a year of research in Egypt, where an extraordinary number of people furthered my efforts. Among them, I would like to thank Samir Farid, Youssef Cherif Rizkallah, and the head of the National Film Center, Madkour Thabet. At Cairo University, Ahmed Zayed proved welcoming, while Aly Yehia, Mona el-Sabban, and Hanna Youssef Hanna at the Higher Institute for Cinema all facilitated my work. Gaber Asfour, the Secretary General of the Supreme Council of Culture, opened doors for me and was particularly hospitable. The late Samir Sarhan, Deputy Minister of Culture, and his assistant, Madame Nagla, enabled me to copy thousands of pages of the early trade journal, *Cine Film*, at the *Dar el-Kutub*. Ali Abu Shadi, the state censor, was open and accommodating, just as Christopher McShane and Nimet Naguib of the U.S. Information Service provided introductions and made calls on my behalf. The staff at Cairo's myriad libraries and state cultural institutions have my permanent gratitude, with Ustaz Ahmad and Madam Maha from the Ministry of Culture's private library among the most obliging in collecting crucial materials.

Filmmakers and industry professionals in New York, Washington, DC, Mexico City, Cairo, and London were unusually generous in helping an interloping political scientist to understand their vocation. Long before making the latest Harry Potter movie, David Yates invited a Yank to the set of his short film, *Punch*—with consequences far exceeding my bit part. In Egypt, Youssef Chahine, Gaby Khoury, Yousry Nasrallah, and Tawfik Saleh were invaluable in bringing Egyptian film to life for me. Bonnie Richardson of the Motion Picture Association in Washington, DC, sent me useful material and put me in touch with General Hussein Hassan Abd el-Rahman of MPA Egypt. My thanks also to Hussein Mutawi, the Director General of the Federation of Egyptian Industries; Antoine and Chady Zeind of United Motion Pictures; Nasser Galal of the Cultural Development Fund; Salah Abd el-Halim Atem of Studio al-Ahram; and Muhammed al-Sayed of Studio Misr.

An additional research trip to Mexico was supported by the Ford Foundation and a Title VI / Foreign Language and Area Studies grant. Several people helped me find my way around Mexico City, including Mario Aguiñaga-Ortuño, the Director of the Cineteca Nacional, and Teresa Zacarías and Bernardo Stril of the state's Mexican Cinema Institute. In addition, Claudia Rivera of Latina films, Ofelia Dominguez de Ruiz of the National Cinema Chamber, and Mariano Sánchez-Ventura of the National Autonomous University of Mexico (UNAM) were particularly generous with their time and insights. The staff at the libraries of the Cineteca, the Colegio de México, the Centro de Capacitación Cinematográfica, and UNAM also furthered my work significantly.

I am grateful to Toby Wahl and the entire team at Palgrave Macmillan, who were exceptionally efficient in handling the manuscript. A very special thanks go to my friend, Jeff Ziarno, for his fine rendering of an original cover illustration. I note that Chapter Three of this book uses material published as "The Globalization of Filmmaking in Latin America and the Middle East," in *The Oxford Handbook of Film and Media Studies*, by permission of Oxford University Press. The American University in Cairo Press was kind enough to allow me to draw on my chapter, "State and Cinema in Pre-Revolutionary Egypt, 1927–52," published in their edited volume, *Re-Envisioning Egypt, 1919–1952*. While I had considerable assistance in Egypt, Mexico, and the United States, I bear full responsibility for the remaining errors of fact or interpretation that will be discernible in a study bridging so many areas to which others have devoted themselves exclusively.

Finally, I would like to acknowledge the influence of my late parents, Rosemary and Joseph Flibbert, who understood that our time spent in *Les Vieilles Vallées* would open my eyes to many other worlds. I dedicate this book to my wife, Sonia Cardenas, and our son, Alex. A fellow political scientist and intellectual collaborator, Sonia's superior analytical talent, editorial wisdom, and unflinching confidence in my work matched her enthusiasm for accompanying me everywhere on this long journey. Our young son, Alex, arrived toward the end but stirred the power of faith and renewal in ways only known by adoring parents.

Note on Transliteration and Translation

Since this book was written for multiple audiences, only one of which is familiar with the Arabic language, I have chosen a simple system of transliteration for the literary Arabic words appearing in some names, movie titles, and source materials. While adhering somewhat to the *International Journal of Middle East Studies* standard, I have omitted all unnecessary diacritical marks, along with most initial *ayns* and *hamzas* in the text, and I have simplified some transliterations to render them as unobtrusive as possible. For well-known personalities and places, moreover, I have used conventional versions, as with Gamal Abd el-Nasser (sometimes shortened to "Nasser"). In cases where the Egyptian colloquial dialect affects standard pronunciations, I have duplicated what will be familiar to most readers, including, for example, the Cairene *geem* rather than *jeem*. With movie titles, I have sought to provide an English translation from the Arabic (or Spanish) original, though just as often I have used the standard translations that are linguistically inaccurate but conventional. I also have provided bracketed translations, usually my own, of important source-material titles to benefit researchers who might not understand the Arabic original.

Part I

The Politics of Cultural Production

1

Introduction

On the night of July 22, 1952, Anwar Sadat went to the movies. A key member of Gamal Abd el-Nasser's clandestine Free Officers movement, he had been stationed at el-Arish airfield in northern Sinai when Nasser called him back to Cairo. Unaware of the exact timing of the impending coup, he went out for the evening with his wife to join other middle-class Egyptians in the country's most popular form of urban entertainment. It was the heyday of the Egyptian cinema, but Sadat and his wife saw an American movie, *High Noon* with Gary Cooper. Technical delays with the film projector kept them late at the open-air cinema on Roda Island, and Sadat returned home around midnight to find a personal note from Nasser that made him blanch. He quickly donned his uniform, grabbed his pistol, and hurried north to the army barracks at Abbasiya. Despite being in the Signal Corps and in charge of communications in the plot, Sadat did not know the secret password needed to get past the barracks guards. His fellow conspirator, Abd el-Hakim Amer, eventually returned from capturing army headquarters, heard Sadat's imploring voice at the gates, and let him in. This allowed Sadat to broadcast the new regime's first communiqué the next morning, maintain his stature among the Free Officers, and one day become vice president and then president of the country. That night's success notwithstanding, he later revealed his angst at having gone to the movies, almost missing the coup that launched the Egyptian revolution.[1]

Sadat's cinematic outing took place more than a half-century ago, but it was years after the establishment of a national film industry in Egypt. While Hollywood had been asserting its global dominance since the 1920s, small producers in places like Egypt, India, and Mexico had emerged with the sound revolution in filmmaking that same decade. They resisted the complete incorporation of their local markets into a globally integrated film trade and built a productive capacity that tapped into audience desires to see and hear culturally resonant films on the silver screen. Just as the globalization of filmmaking began decades ago, so did resistance to such pressures, even if it was partial, half-hearted, and sometimes short-lived. Nationalist coup plotters like Anwar Sadat might have worn Western-style military uniforms, espoused secular ideologies, and enjoyed allegorical Hollywood westerns, but often they found diversion and stimulation in their own, locally rooted forms of cultural expression. In this manner, those who experienced globalization's early, sharper edges sometimes defied and contested its apparent seamlessness and inexorability.

Nonetheless, the conventional wisdom today is that globalization creates irresistible incentives for states to liberalize their control over goods, services, capital, information, and ideas. Observers have either hailed or lamented a presumed worldwide economic and cultural convergence, as liberal and homogenizing pressures sweep the globe and transform the institutional landscape. Such pressures and associated policy changes have profound implications for national sovereignty, economic governance, and cultural autonomy. Since globalization erases boundaries, increasingly dense international connections require even large states to adjust their policies in a wide range of areas. State policy makers may retain some freedom of choice, but the structural consequences of global integration are difficult to ignore.

This reasoning is commonly extended to state policy in matters of national culture and identity. When there are serious financial costs to maintaining cultural autonomy or defending expressions of national identity, states are expected to sacrifice the latter concerns for the sake of compelling economic imperatives. By this logic, a failure to permit free trade and open markets in culturally significant commodities imposes unsustainable long-term costs, so that economic necessities eventually overwhelm all others. The economically marginal cannot afford the luxury of local film industries when globally oriented producers are available to meet their needs. This marks the final victory of late capitalism in an international economic environment defined by the triumph of the market and the unsustainable nature of the alternatives.[2]

The seeming futility of resistance notwithstanding, some states in the past century have resisted international pressures to liberalize trade in goods that are deemed to have national cultural significance, such as filmmaking, publishing, music, photography, and certain foods and textiles. They have done so even when resistance is economically irrational and punishingly expensive to consumers and state budgets. At a minimum, state policy makers have responded in widely varied ways to the fiercest international competition, both historically and cross-nationally. Responses have ranged from full liberalization in some cases to enhanced protectionism in others. Belying the expectations of more breathless accounts of globalization, states have reacted in surprisingly inconsistent and contradictory ways. State leaders have not had a free hand in their decision making, though they usually have paid close attention to cultural production as it intersects with nationalist endeavors. This story has unfolded both in distant locales like Egypt and India, as well as in more culturally and physically proximate places like Mexico.

Mexican state officials, for their part, have usually taken filmmaking seriously, mostly because the local industry was born just as Mexican nationalism was being redefined after the early twentieth-century revolution. They founded the world's first film archive in 1936, debated the industry's merits on the Senate floor in 1937, and helped to establish the world's first film-financing cinema bank in 1942. A long succession of Mexican presidents promoted filmmaking as integral to the international projection of Mexican national identity, implored the industry to improve its sinking quality, and supported it long after it had become stale and decrepit. President Echeverría was so committed that he all but nationalized the industry in the early 1970s, just as the world film trade was entering its most fully integrative period. He appointed his brother—a film director—to head the cinema bank and sent the cinema bank president to Washington as Mexico's ambassador to the United States. Even subsequent efforts to cut the industry down to size received surprisingly high-level attention, with President López Portillo choosing his sister in the 1980s to oversee the dismantling of a powerful public-sector film apparatus that had endured for decades.

Given the high level of attention to cultural production and trade in Mexico, Egypt, and beyond, how does one situate smaller, less-powerful culture producers in the long engagement of a century of globalization? Why, for that matter, do states vary so much in their responses to the competitive pressures of globalization? Why do some states accept economic losses in exchange for cultural gains? How much, if at all, do states still matter, and how much autonomous control do they retain in the cultural realm? What factors shape the production and marketing of cultural commodities in the world economy? Who bears the costs of cultural protectionism when such policies are chosen, and who reaps the benefits? What, finally, is the relationship between the economic and cultural dimensions of globalization?

Policy Responses in the Film Trade

For the past several decades, the world film trade has been transformed by the growth and development of integrated international markets, which have raised the level of competition for local producers everywhere. Major producers began exporting film early in the twentieth century, and since the 1960s, leading firms—mostly American in origin—have viewed foreign markets as equally important to domestic ones, obtaining roughly half their total revenues abroad. Global integration has affected all film producers. Companies in the smallest film-producing states now compete locally with major, internationally integrated firms, while midsized, regionally dominant film producers have seen the incorporation of their traditional markets into world trade. Such changes have curtailed local exports and rendered industries in some countries nearly inviable, as shrinking market share has reduced profits and depleted the amount of investment capital available for continued production.

Film-producing states have responded to international pressures in highly varied ways, both cross-nationally among the world's many producers and over time within the same ones. One of the strongest dimensions of this variation is in the extent to which film producers are open or closed to exports from powerful countries like the United States. Protectionist measures have not been implemented uniformly throughout the world in response to American-led global competition, and a great many states have maintained liberal policies toward the film trade. Since the end of World War I, American film exports have made their way nearly everywhere, and the United States has managed to gain the lion's share of film revenues in most of the world's markets. While the major U.S. companies have experienced periodic reductions in their access to particular places, their total share of the foreign market has remained extraordinarily high. At times, even the largest film-producing countries have imported a great many American films, with the reduction of trade restrictions prompting cries of cultural imperialism by critics of American influence and proponents of national culture.

Nonetheless, protectionism in the film trade has been common and substantial, even if such policies have been costly to consumers and state budgets supporting local industries. Since the 1920s, European producers in Germany, France, Italy, and Britain have contested U.S. dominance and sought protection for their industries from American competitors. Perhaps more surprisingly, relatively weak non-European producers in Africa, Latin America, Asia, and the Middle East also have protected national film industries in the face of the growing integration of the world film trade. When and where film producers have obtained it, protectionism has taken the form of an unusual variety of trade barriers, including tariffs, quotas, discriminatory taxes, licensing requirements, and financial restrictions, as well as more indirect means like subsidies for domestic producers and weak enforcement of intellectual property laws. At times, policy makers in most states have worked actively to regulate, to promote, and to shape the industry to suit larger national purposes, treating it quite differently from many other sectors. Confronted with the steadily growing competitive pressures of globalization, they have chosen to sustain film industries that have great difficulty competing with their American rivals.

In short, comparable levels of international competitive pressure have yielded very different responses in places that otherwise are similarly situated in the industry. The source of this variation is not self-evident, nor is it explained easily by conventional accounts in political economy. Variation has not been simply a function of state power, regime type, domestic political coalition, or level of economic development. In the immediate post–World War II era, for example, countries as different as Germany, Egypt, and the United States had relatively few restrictions, while comparably distinctive producers like France, Mexico, and India erected substantial barriers to film imports. In no case has film-related trade policy anywhere remained unchanged over the years: the level of protectionism in markets like Britain's has oscillated dramatically from extremely high—with import duties of 75 percent in the postwar era—to virtually nonexistent.[3] In responding to similar levels of international pressure from the United States, the policies of key regional producers like Egypt and Mexico have moved in reverse directions: Egypt had a liberal orientation in the 1930s and gradually became more protectionist, while Mexico started with protectionist policies and liberalized over time.

What explains this variation? If competition has grown steadily since the early years of filmmaking, why have world markets in the sector not become uniformly more protectionist? If globalization is as irresistible and overpowering as many claim, why have state policies not become uniformly more liberal? What are the most influential sources of policy change in this area: economic interests, cultural concerns, or a political logic more fundamental than both? What factors, for example, are at the heart of French-led resistance to liberalization in the cultural industries? Is the film trade any different from that of other sectors, and if so, how and why?

Back to Domestic Politics

Prominent domestic differences among states generate a diversity of responses to globalization in filmmaking, and these differences have fundamentally political origins that persist in the face of global market pressures. Such differences are not reducible to the six-year Mexican presidential *sexenio*, nor are they merely a function of the changing whims of a succession of Egyptian state leaders over the decades. As I demonstrate in this study,

Figure 1.1 State Responses to Globalization

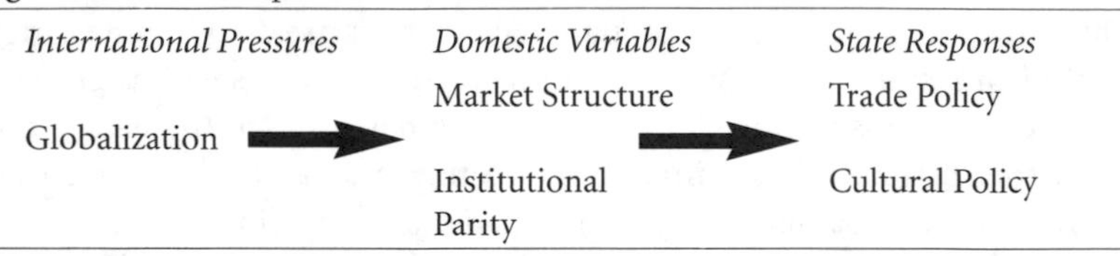

two elements take precedent in this regard: market structure and institutional parity. Market structure refers to the degree of domestic competition in an industry, and it affects dramatically the capacity of societal actors to promote common interests, shaping trade policy in cultural products. Institutional parity speaks to the relationship between economic and cultural institutions in the film sector, and it determines the extent to which cultural concerns are given voice in policy outcomes. Market structure and institutional parity account, respectively, for the trade and cultural dimensions of state policy (see Figure 1.1).

Political scientists have highlighted the role of domestic politics in constraining the effects of globalization, but I extend this work in two ways. First, I problematize the concept of market structure to show the real-world challenge of mounting collective action in response to globalization pressures. Globalization may shape the preferences of societal groups, as many have argued. However, the capacity of these groups to organize—and therefore to influence state policy—is affected by the structure of the domestic market in which they operate. Second, I introduce the concept of institutional parity to capture the relationship between contending state institutions with overlapping bureaucratic concerns. Institutional parity is significant because it shapes the basic interests of state decision makers, affecting whether they use cultural or economic criteria in policy making.

The argument moves beyond accounts that explain national responses to globalization in terms of specific societal actors or state institutions. Rather than examining individual groups or agencies, I focus on the ownership and organizational links between, first, the industry's various subsectors (market structure) and, second, the rival state institutions overseeing policy making (institutional parity).[4] In so doing, I contribute to the growing literature on institutions, while expanding the scope of debate about the many faces of globalization. In a world where economic and cultural issues are linked inextricably, this is an effort to illuminate some of the connections between one set of such issues while showing exactly which state institutions matter, and why.

Substantively, moreover, this study challenges the traditional boundaries of trade policy analysis by examining an industry in the service sector, which until now has been neglected by political economists. This sector includes media and information industries that have captured a rapidly growing percentage of international trade in recent years, and that are a leading and consequential part of the new economy in the United States and elsewhere. This represents a substantial corrective to the outdated, sometimes Marxist-inspired, emphasis on manufacturing and industrial production, which has blinded contemporary analysts to the multifaceted significance of the trade in services.

I examine the evidence primarily through a comparison of Egypt and Mexico, as well as by considering the role of the United States in their respective regional markets. Hollywood's longstanding prominence in the world film trade is well known, while Egypt and Mexico historically have been leading regional film producers in the global South. The book therefore addresses both cultural flows and developmental challenges, incorporating them into a research agenda that centers typically on the economic travails of the global North. This is important in itself, because the movement of cultural commodities across state borders has been a longstanding concern to policy makers in all corners of the world, whether in metropolitan centers like France and the United States or in the more distant economic and cultural periphery.

Why the Film Trade?

Situated at the nexus of international commerce and popular culture, film is both a major economic commodity and one of the past century's most important forms of cultural production. In several different ways, film's complex and multifaceted nature enhances its significance as a means of understanding larger changes in the world. First, the economic characteristics of commercial filmmaking create powerful incentives to export, making the industry a useful case study of domestic-international interaction in political economy. Distinct nationally based film industries began competing internationally within a decade of the invention of the first commercial projector in 1896. While the national affiliations of producers have since become much more complex, the industry remains subject to both domestic and international imperatives. Neither level of analysis can reasonably be excluded from a study of outcomes in this area.[5]

Second, the film industry serves as a window onto multiple aspects of the globalization debate, since the multifaceted significance of film speaks to many different kinds of changes occurring in the world.[6] At present, a clear divide exists between studies that address the economic dimensions of globalization and those that focus on its cultural aspects, with the crucial links between them remaining vastly underspecified.[7] The film industry is neither simple commercial activity nor pure cultural expression, given the semiotic dimension of the most unapologetically commercial production and the financial requirements of even low-budget art film.[8] Film production and trade encompass both sides of the debate and may speak to us equally well about the transformation of the world economy and the politics of cultural change. Just as the pressures of globalization forge international connections, a study of this industry serves to build intellectual connections across related areas of research.

Third, most studies of trade in political economy remain focused on manufactured goods, with insufficient attention being devoted to services. No study, moreover, has employed the film industry to address this increasingly important area. While the film product is itself a tangible commodity with physical attributes, filmmaking conventionally is classified by economists as part of the service sector because of the manner in which it is "sold" at the retail level: distributors sell limited exhibition rights to theaters, which charge "rents" to moviegoers for a one-time viewing opportunity.

As an economic activity, the film sector shares important similarities with other services, such as engineering, banking, advertising, consulting, and tourism. In its purely economic characteristics, for example, making movies is more comparable to writing new software than to producing automobiles: all production costs go toward a single first print of the film. This confers greater risk on investors and necessitates, as with other consumer services, buying and selling without the physical inspection and transfer of commodities.[9]

The growth of services has had clear political consequences that are evident in stronger international efforts to oversee these activities, most immediately in the creation of new international institutions to govern the service trade. As services have risen dramatically in economic importance compared with agriculture, mining, and manufacturing, the regulation of this sector has been codified in the General Agreement on Trade in Services (GATS) and the Agreement on Trade-Related Intellectual Property Rights (TRIPS).[10] Yet with few exceptions, most theories in political economy treat goods and services the same, despite the distinctive dynamics of commodities with strategic or cultural significance. Such assumptions derive from the requirements of economic theory, but changing political and economic realities may demand scrutiny of the special attributes of the service sector.

Fourth, the industry itself has long been characterized by extensive internationalization, with some degree of integration occurring on several levels, including in markets, financing, the technical aspects of production, and skilled labor.[11] The relatively low cost-to-profit ratios for transporting film across the world have meant that the industry never was as constrained in exports as industries producing bulkier goods. With the low costs and high benefits of international trade in the sector, industry leaders had little choice but to internationalize very early on. For this reason, the U.S. film industry began underwriting the integration of world markets by the end of World War I, when the film trade was still in its infancy. Accordingly, film markets for most major producers never were segmented fully by national borders. This long record of internationalization in filmmaking allows for an examination of continuity and change in the effects of national structures of production and regulation. It also serves as a bellwether of future trends in other heavily internationalized industries, for which the experience of the film sector may offer significant lessons.

Finally, in contrast to most other major industries, the political dimension of film exports—like those of other cultural commodities—provokes contestation and heightens their potential significance for domestic and international politics. This is no less true of ordinary, commercially oriented melodramas, which use language and cultural referents to convey ideas and therefore cannot have the relative ideological neutrality of more prosaic goods. Even among those countries that share common languages or have cultural affinities, policy conflict has been severe at times. Trade in cultural products, for example, was hotly contested in the final round of negotiations for the General Agreement on Tariffs and Trade (GATT), concluded in April 1994. French demands for a "cultural exception" for subsidies and tariffs protecting its film industry nearly undermined the entire agreement. France's eleventh-hour success only highlights the salience of cultural products for future discussions under the auspices of the World Trade Organization. Efforts to keep film-related legislation out of larger trade liberalization agreements are a point of impending contention, as key states resist policy harmonization.[12]

Beyond Systemic Explanation

Culture never before has been so fully subject to the logic of the market, but specialists in the politics of international trade have rarely addressed cultural products in general or the film industry in particular. A number of valuable studies of film and politics have been written from historical and interpretive perspectives, primarily considering issues of early trade relations, film substance, national identity, cultural imperialism, or the emergence of cinematic alternatives to Hollywood.[13] Yet very little has been written from the theoretical or methodological viewpoint of the social sciences, least of all involving a self-conscious effort at hypothesis testing and theory-building. This lacuna is important because none of the existing studies offers a satisfactory explanation for the extraordinary range of state responses to globalization pressures.

Theoretical work in international political economy provides three long-standing systemic approaches that might claim applicability to this area of research: realism, liberalism, and dependency theory. According to each of these perspectives, the international system strongly circumscribes policy choices by influencing the set of viable options that states have at any given time. Foreign economic policy is said to result substantially from factors that are exogenous to individual states, beyond their control. Consequently, state policy makers commonly are treated as either relatively powerless to resist compelling external forces (realism), obstructionist and unwise when they do so (liberalism), or complicit in thwarting genuine development (dependency theory). I do not challenge these theories per se, particularly in their usefulness as first cuts at explanation, but I do join those who highlight the limits of third-image theorizing in accounting for varied national responses to globalization pressures.

Emphasizing international structural factors, neorealism stresses the importance of relative power in the interstate rivalry it claims is inevitable under conditions of international anarchy. Realist theory in general suggests that trade-policy choices generally reflect the pursuit of relative power gains by rational unitary states, with more powerful states constraining the choices available to weaker ones. From this viewpoint, states sometimes adopt economically suboptimal protectionist policies to further their security by preventing adversaries from benefiting disproportionately from a trade relationship. With security concerns necessarily paramount in the hierarchy of state interests, policy in all other areas—especially for "softer" issues in the economic and cultural domain—is secondary to maintaining relative power.

According to one variant of realism, a hegemonic state works actively to maintain a stable and open international economic order, because it is willing to bear the costs of sustaining such a collective good.[14] This view is echoed, if unintentionally, in the most influential historical accounts of the U.S. film industry's longtime dominance of the world film trade, especially regarding the control of world distribution networks.[15] American political and economic power is claimed to have bolstered Hollywood's hegemony and supported the growth and integration of world film markets, with the United States profiting greatly by virtue of its large film output. From this perspective, the U.S. industry has endeavored to guarantee free trade in film for several decades, securing official intervention on its behalf and deploying its own considerable power to punish even the smallest attempts at closure. If extended explicitly, such a theory would predict that film-related economic policies, in general, should covary with the ascendance or decline of powerful states. A film-industry hegemon, by this logic, should be able to control its most important markets and assure itself key export outlets. Declining hegemonic power, realism predicts, elicits rising policy defection, and greater protectionism by smaller, weaker states.

A review of the evidence, however, shows that the American film industry has not been able to control the world film trade to the extent expected by a state-power perspective. Resistance to U.S. film hegemony has been substantial, and general policy patterns have not varied simply with changes in American power. While the U.S. government has intervened diplomatically on behalf of the film industry, this intervention has been uneven and at times quite unsuccessful.[16] More broadly, a realist explanation for larger patterns in the international political economy cannot account easily for differences in states' foreign economic policies. Despite extensive changes in the policies of most major film-producing countries, there has been little variation in their relative power internationally. Surely, state power considerations have played a substantial role in the U.S. film industry's success, but the simple fact of American dominance does not explain the variation of responses to global pressures by other states.

A competing liberal approach sees international competition in a more benevolent light. This view derives partly from the economics and public choice perspectives and was echoed by U.S. trade negotiators in the final stages of the GATT talks. It holds that rational, self-regarding states will pursue trade policies to achieve absolute gains in all areas, creating multilateral institutional mechanisms to solve the problems of cooperation. Sound policy choices are to be based on the promotion of economic efficiency and the pursuit of comparative advantage. The United States, for example, is said to have a comparative advantage in filmmaking, since it is capital rich, and the vast size of its domestic market enables producers to invest in expensively made movies, amortize costs in the home market, and sell abroad at discount prices. Like neoclassical trade theory, a liberal approach therefore considers any protectionism in the film sector to be economically myopic on a national level.[17] This view expects protectionist policies only when cooperation-inducing international institutions fail and the firms and politicians who would benefit from rent seeking are positioned to obtain it by political intervention. Typically, liberalism predicts, poorly endowed film producers will lobby for protection from international market forces. In so doing, they limit the choices and quality of films available to domestic moviegoers in an effort to prop up inefficient, uncompetitive industries.[18]

While many of the world's cultural producers have not competed effectively with their counterparts in the United States, a liberal explanation for trade policies in this sector is no more compelling than that of realism. An efficiency-based criticism of protectionism is built on the assumption of rent seeking as the leading cause of protectionist policies. This view is persuasive in some contexts, but it cannot explain the full range of cross-national variation or policy change over time. Most significantly, it underestimates the possibility that state authorities themselves may prefer policies that are economically costly on a national level, but politically beneficial in ways other than as rent seeking. By assuming that state policy toward cultural products normally will be based on market criteria, or that policy makers will accept the "natural" advantages conferred by factor endowments or market size, a liberal approach offers only a limited explanation of state responses to global economic and cultural pressures.

Finally, accounts derived from dependency theory emphasize the manner in which multinational corporations control global film markets and undermine local production, often through an alliance between foreign distributors and local exhibitors.[19] Sometimes invoking either Immanuel Wallerstein's related world-system theory or an earlier "Third Cinema" movement in filmmaking, this approach attributes much of the weakness of non-Western film industries to the economic and cultural imperialism of the United States via Hollywood.[20] The newest version of the thesis manifests itself in an overtly normative condemnation of the economic and cultural effects of globalization in the developing world. The post–cold war era, from this perspective, is increasingly defined by the homogenizing effects of a global economic and cultural order dominated by the United States. Advocating import substitution as a strategy to break free from dependency and the perils of globalization, this approach expects protectionist policies to be most prevalent when comprador elites are weakest and in sectors that lend themselves most easily to import substitution.

Yet this view starts from a problematic set of assumptions, including that a state's position in the international economy determines its policy choices, that states are relatively helpless in responding to the pernicious effects of global economic forces, and that integration inevitably entails losses that are avoidable by pursuing more autarkic policies. While, no doubt, specific forms of dependency have existed in filmmaking and other cultural industries, it is not clear that such dependency is one sided, given the remarkable degree to which the American film industry has come to rely on foreign markets. Smaller producers may actually suffer greater harm from failed exports and domestic price distortions than from excess imports, especially in segmented markets with room for both international and local production.[21] Perhaps most importantly, such a rigidly structuralist position cannot account for a common occurrence: differences in the policy choices of states that are similarly situated vis-à-vis the industrial "core." International constraints are real, but by failing to take into account such diverse domestic contexts, little contribution is made to understanding the sources of policy variation.

The foregoing is not to deny the value of realism's emphasis on power considerations, liberalism's focus on international institutions, or dependency theory's highlighting of structural inequalities. Admittedly, sophisticated versions of all three theories could be used to explain important global trends. Yet by obscuring the extent to which state policies vary, even across states that are similarly situated internationally, these perspectives have inadequate theoretical leverage to explain the outcomes that are central to this book.

The Scope of the Study

Studying a single industry has distinct advantages, though a narrow focus imposes inevitable limitations on research findings and generalizability. The principal advantage is that it controls for the sectoral context, which is a possible explanation for variations in national responses to globalization. I minimize the associated dangers by deploying a rigorous and multitiered comparison. In addition to cross-national variation, I explain the varied consequences of globalization over time and across different subsectors (i.e., film production, distribution, and exhibition). That said, this study does not offer a general explanation for all trade and cultural policy outcomes, or for the preferences of all state and societal actors vis-à-vis globalization. Nor does the book pretend to be an exhaustive analysis of filmmaking in its economic or artistic dimensions, a task better left to researchers in other fields. Certainly, specialists in any of the cases I consider will find specific points of fact or interpretation with which to quibble, but my purpose is to put the industry in the service of explaining broader political problems, rather than to treat it as the exclusive subject of inquiry.

To maintain some theoretical focus, moreover, I am compelled to omit detailed consideration of related service industries, including the parallel emergence and development of television, video, cable, and other delivery media for cultural goods. The latter have gained an increasingly substantial share of the total market and have their own significance for the questions I address. Nevertheless, the emphasis here is on the production and trade of feature-length films, which have the longest history of political conflict surrounding them, arguably the most generalizable economic characteristics, and are the most readily distinguishable cultural goods available for study. While the argument may be suggestive of dynamics that are applicable to newer media like the Internet, the rapidly evolving nature and unique attributes of this technology render it less useful to a study of the political economy of cultural production.

Data limitations also present special challenges to researchers in this field, affecting the kinds of assessments that can be made. This is doubly true for the Egyptian and Mexican cases, where the most basic economic data—if at all available—are notoriously unreliable. To counteract these weaknesses, I have conducted supplementary personal interviews, but these have some limitations as well. The most knowledgeable industry insiders are either businesspeople who consider their economic insights to be comparable to trade secrets, or creative and artistic personnel who often have less enthusiasm for the business-related aspects of a rigorous empirical study. Still, these problems are mitigated by the relative ease of estimating the direction of bias in published figures and the assessments offered by industry professionals. Aggregating data also reduces much of the uncertainty regarding policy outcomes, which can be gauged most reliably in general terms. The historical scope of the work, while presenting its own difficulties in data availability, likewise contributes to confidence in the findings, since a broad view of the contours of the film sector is less likely to hinge on the reliability of data in any single period.

The remainder of the book is organized as follows. Chapter 2 elaborates the argument, a set of claims, and a methodology for evaluating them. Chapter 3 discusses the international dynamics of the world film trade, showing how the changing level of international competitive pressure affects each case but is inadequate to explain the full range of policy outcomes. Consequently, I introduce the domestic level of analysis in Chapter 4, demonstrating how market structures shape trade policy in cultural products. Chapter 5 then examines the role of state institutions in accounting for cultural policy, highlighting their effects on the rivalry between economic and cultural interests. Finally, Chapter 6 offers broader conclusions and discusses the implications of the argument for explaining policy responses to globalization pressures that cross the economic and cultural divide.

2

Explaining Policy Choice

Even in a globalizing world, states sometimes forgo international commerce for national culture, sacrificing economic opportunities to advance or defend cultural goals. This chapter offers an explanation for this trade-off. The starting point of my account is the well-established contention that domestic politics have a substantial influence on state responses to external competition. Globalization creates powerful international pressures, but they are filtered through the prism of national political institutions and economic structures. Even over long periods of time, the latter often prove resilient, continuing to exert independent influence. The indeterminacy of international pressures therefore requires a comparative analysis of the diverse contexts in which the seemingly irresistible power of globalization meets, and is mediated by, local influences.[1]

State responses to a globalizing environment in filmmaking are expressed through both trade and cultural policy. The regulation of film necessarily involves cultural as well as economic choices, since it affects public life in both domains and may be subject to their contending imperatives. "Cultural commodities" and "cultural products" embody the dual nature of the goods and services that state policy makers sometimes use to shape national identity. State policy toward cultural commodities does not always differ from policy toward other kinds of goods. The public aspect of film, however, means that the narrow logic of economic profit sometimes finds itself at loggerheads with the broader logic of cultural promotion and identity-building.[2]

Responses to Globalization

While the systemic theories described in Chapter 1 cannot easily explain variations in state responses to globalization, alternative approaches incorporate domestic variables more fully into their analyses. By specifying the connections between the domestic and international levels of analysis, these accounts have greater explanatory power than their systemic counterparts. In particular, leading approaches in the literature have employed domestic political factors to show how international competition shapes societal preferences, and to account for how such pressures are mediated by variables in the domestic institutional context. I discuss two of these perspectives very briefly before turning to my own argument, which builds on this earlier work.

The "second image reversed" tradition emerged in the late 1970s to show the insufficiently acknowledged impact of the international economy on domestic politics.[3] Much of this work focuses on the role of societal actors as diverse as industrial sectors, individual firms, or factors of production. Subsequent research has extended this line of inquiry to address the influence of international economic conditions on the policy preferences of interest groups, the effects of export dependence and multinationality on the trade preferences of firms, and the consequences of shifting trade patterns for domestic political cleavages.[4] Its key contribution includes the specification of a theory of domestic preference formation that is rooted in international variables.

A second and closely related literature focuses more explicitly on the many ways in which domestic institutional contexts mediate the impact of external pressures. It demonstrates, for example, how the partisan composition of government, the organization of labor, and the strength of political institutions may affect state responses to international economic pressures.[5] Associated in some cases with the revival of institutional analysis that began in the mid-1980s, this work sees policy differences as resulting from domestic institutional variation. One of its most important contributions is to show how variations in the organizational capacities of societal actors help to translate domestic interests into policy outcomes.

While both approaches explain certain aspects of national responses to external pressures, neither addresses the particular interconnections between domestic actors that are acutely significant in a globalizing context. Existing work, for example, often gives analytical primacy to classes, firms, or factors of production. Yet in a world of growing global connections, significant industrial cleavages and alliances are occurring at the subsectoral level. State borders may be diminishing in importance, only to be replaced by new divisions within sectors themselves.[6] Likewise, while analysts often treat political institutions in isolation from one another, the nature of relationships between and among institutions matters as well. This is especially true when such institutions represent different kinds of interests vying for expression in state policy.[7]

The Argument in Brief

Explaining responses to globalization requires determining how different configurations of state-society relations mediate international competition. Surely, the global integration of the film trade has increased competition for industries everywhere. But the heart of the globalization debate concerns the effects of such rising pressure, and what remains an open question: can, and should, the state limit the negative consequences of foreign competition? In practice, state responses are shaped by two domestic factors—market structure and institutional parity—reflecting a particular historical trajectory and influencing trade and cultural policy accordingly.

In most general terms, societal capacity to gain protection for any given economic sector is greatest when the domestic industry is organized noncompetitively. In other words, monopolies tend to need, seek, and win protection. In addition, when trade pressures are directed at countries with state cultural institutions that enjoy a relative structural parity with economic ones, "promotive" cultural policies usually ensue, even if this is economically costly. Put more simply, states promote cultural interests when their cultural institutions are on par with their economic ones. The signal example of the latter pair of outcomes—encompassing both societal and state factors—is when a culture-producing monopoly allies with a powerful and independent Ministry of Culture (Table 2.1, quadrant I).

Alternatively, when an industry is organized more competitively, it will seek a more liberal response to trade pressure. When these demands are

Table 2.1 Trade and Cultural Policy in the Context of Globalization

		Market Structure	
		Monopolistic	Competitive
Institutional Parity	High	I LEAST GLOBAL Protectionist trade policy Promotive cultural policy	III ECONOMICALLY GLOBAL Liberal trade policy Promotive cultural policy
	Low	II CULTURALLY GLOBAL Protectionist trade policy Laissez-faire cultural policy	IV MOST GLOBAL Liberal trade policy Laissez-faire cultural policy

made of countries with state cultural institutions that are not endowed with a relative parity to economic ones, the most liberal responses will tend to follow (Table 2.1, quadrant IV). Without an institutionalized, noneconomic rationale for protecting film industries, and without the unity of interest and political power that comes with oligopoly or monopoly, pro-protectionist groups generally are too fragmented to realize their ends. This logic usually holds even when liberal policies prove destructive, undermining the capacity of poorly endowed industries to survive global competition. These outcomes are depicted in very general terms for selected film producers in Table 2.2. A more detailed discussion of the argument's three principal elements follows.

International Competition

The degree of international competition is an exogenous, generally uncontrollable factor to which all policy makers must respond. Whether in the form of foreign imports in the domestic market or rival exports in foreign markets, international competition pressures local industries, posing challenges to individual firms and states. Firms experience these pressures through declining shares of the market. State policy makers feel them in two ways: through the impact of competition on domestic industry, which usually translates into a political response from the sectors affected; and

Table 2.2 Film Policies for Selected National Producers

Cultural Policy	Trade Policy: More Protectionist	Trade Policy: More Liberal
More Promotive	I France 1990s Egypt 1960s Mexico 1940s U.S.S.R. 1950s	III Italy 1950s Britain 1980s Singapore 1990s Netherlands 1960s
More Lassiez faire	II Egypt 1980s India 1980s Japan 1970s U.S. 1940s	IV U.S. 1990s Mexico 1990s Egypt 1940s Germany 1950s

by their effects on national concerns, ranging potentially from the balance of payments to political conflict over cultural identity. The distinction between firms and states becomes blurred under conditions of state ownership of the sector, but trade pressure always originates from a political and economic source that is beyond the immediate control of any one firm or state.

The most substantial and persistent competition in the world film trade has come from a small number of powerful, globally integrated firms of U.S. origin, particularly the "majors" that were established early in the twentieth century. American hegemony in film dates to the sudden drop in French production during World War I, after which U.S. dominance extended worldwide and increased the level of competition for all producers. Subsequent developments in the American industry have ramified throughout foreign markets, driving key changes elsewhere and shaping the opportunities and constraints facing other film-producing countries.

Three significant periods of film trade are discernible, each showing the degree of international pressure and its effects. First, after an initial struggle for control over crucial patents and production technology, a consolidated, vertically integrated oligopoly ruled Hollywood in the 1930s–1940s. This was the American studio system's heyday, and its extraordinary domestic success facilitated the worldwide establishment of the American film as the defining form of the medium. A second period in the 1950s–1960s began with a series of decisive domestic challenges to the U.S. industry, including the landmark Paramount antitrust decision, postwar demographic changes, and the rise of television. These challenges provided new incentives for the American industry's further expansion into overseas markets in an attempt to compensate for declining domestic returns. Finally, the 1970s–1990s saw the conglomeration of the industry, when leading U.S. firms sought relief from the high costs and risks of filmmaking by joining either larger entertainment interests or more diversified multinational business enterprises. As a result, American dominance of the film trade became linked to the globally integrated firms so characteristic of the contemporary economy, and production was dispersed throughout the world under a system of flexible specialization that further increased the competition faced by most national producers.

Changes in global markets in each period affected the policies followed by other states engaged in the film trade, none more so than the small-country producers of the global South. Indigenous film production began in many developing countries with the advent of sound recording in filmmaking in the late 1920s, which enabled producers to make movies in local or national languages. National cinema industries were built, for example, in Egypt, India, and Mexico in the 1930s, as local entrepreneurs, sometimes aided by the state, gained access to the new technology. As filmmakers in the latter three countries established themselves in the regional markets of the Middle East, South Asia, and Latin America, they competed for market share with a larger U.S. industry that they tended to emulate. At times, this led to conflict between the smaller, regionally oriented producers and their American competitors, spurring efforts to obtain protection from U.S. import pressures.[8]

Significantly, competition between regional and global film producers is not always zero-sum in nature, leading some observers to conclude that trade pressure is rather inconsequential. In some markets and at certain times, foreign- and domestically made films do complement each other, especially where domestic production levels are insufficient to meet the needs of exhibitors, who require a steady supply of movies to fill theaters and generate predictable revenues. In economic terms, moreover, film products are never perfect substitutes: audiences may wish to see an expensively made or exotic foreign movie one night and a cheaper but more locally resonant one the next. Finally, high-quality films made in one industry can serve as a stimulus to consumer demand for all films across the board, as the moviegoing habit is developed and incorporated into weekly social routines.

These potentially mitigating factors notwithstanding, international competition between regionally and globally integrated producers still has mattered greatly for industry and policy-making dynamics. No doubt, producers in most developing countries have never been able to make enough films to meet audience demand and therefore have acquiesced to imports. Yet the leading producers in the industrialized economies faced this problem only in their early years, before other countries' production facilities were fully developed. Anytime thereafter, whenever production by smaller countries was insufficient, it was due largely to the dearth of investment capital and the shrinking of already small market shares that resulted from foreign competition. Furthermore, claims for the nonsubstitutability of film are entirely valid, but they represent the consumer's perspective, not that of the producers, distributors, and exhibitors who feel competitive pressures and influence trade policy. Film consumers themselves are much more likely to express their film-related preferences at the box office than in the ballot box, given the high elasticity of demand for movies. Finally, consumer demand stimulated by foreign films does not lessen trade pressure, since it may only heighten moviegoers' preferences for foreign imports.

If international pressures are sufficiently important by themselves, then any change in the level of trade pressure for a particular market should lead to policy changes. Specifically, greater (or lesser) international pressures should induce policies favoring greater (or lesser) closure. By this logic, the shift in the world film trade that began around 1950, from Hollywood-dominated oligopoly to crisis-induced foreign expansion on the part of U.S. firms, should have yielded greater protectionism in film markets worldwide. Oddly enough, it did not. Again, toward the end of the 1960s, the industry's nascent globalization should have elicited further efforts by local industries to protect their markets. They did no such thing—at least not uniformly. This logic should hold both internationally and regionally: as regional markets expand in importance to global firms, local protectionism should follow, with trade competition between producers playing out in third countries. However, since changes in regional and international competition do not elicit automatic, uniform responses, the mediation of domestic variables must be incorporated into the account.

Market Structure

Film industries always serve relatively large, diffusely organized groups of individual consumers, who are prevented by collective action problems from controlling the policy-making process.[9] Groups divided by otherwise important social cleavages, such as class, ethnicity, or factor of production, do not tend to cohere around filmmaking to engage in industry-related activism. For this reason, the most significant societal actors in commercial filmmaking are the industry's firms and workers, grouped together in the three major subsectors of production, distribution, and exhibition.[10] Regardless of their artistic commitment to filmmaking, the film sector's firms and workers seek generally to maximize revenues and wages. This is not to say that noneconomic motives are absent, but those engaged in filmmaking on a professional basis tend to behave like other necessarily self-interested economic actors. Accordingly, the responses of firms and workers to external competition are a function of market structure, or the degree of domestic competition that divides or unites them in action and interest.

Market structure refers to the number of existing firms, the ease of entry for new firms, and the number of consumers in a given sector.[11] This variable embodies the capacity of societal actors to articulate coherent demands and lobby policy makers for their preferences. It also indicates the compatibility of interests for the various subgroups in the sector. If the industry is concentrated in few firms, is difficult to enter for new ones, and serves numerous consumers, then firms and workers can demand more effectively their preferred policies. With significant interconnections among firms, their interests tend to be more compatible, as they are better positioned to work toward centrally defined goals. In its purest capitalist form of industrial monopoly, such a market structure is relatively close to noncapitalist, hierarchical means of allocating resources, as in the state-owned enterprises of command economies. Regardless of who controls ownership, if protectionism induces monopoly—a long-established principle in trade theory—then monopolists of any stripe should seek protection when their share of the market is threatened from without.[12]

By contrast, the more competitive the market structure—with many firms, easy entry, and few consumers—the harder it is for sectoral actors to mount collective action in support of their policy preferences. Without, moreover, a centralized administrative mechanism to coordinate efforts, firm and worker preferences are likely to be relatively incompatible, since differently positioned subsectors may benefit from different policies. The latter model in its purest form, with many small and unconnected firms, approaches the neoclassical ideal of the perfectly competitive market.[13] Markets, like other institutions, are the product of historical processes that have varied over time and from place to place. The variable of market structure therefore serves to problematize the existence of self-regulating markets as simple price-generating mechanisms, revealing their contingent nature and their differential consequences for political action.[14]

In the film industry, variations in market structure have clear consequences for the capacity of social actors to organize within and across the distinctive subsectors of production, distribution, and exhibition. Without connections of ownership linking them, firms and workers in these subsectors have a hard time cooperating and are likely to hold sharply conflicting policy preferences. As Figure 2.1 shows, such preferences can be read off of subsector positions in the industry: Producers usually have the least extensive and direct ties to foreign markets, giving them strongly protectionist preferences when confronted with trade pressure.[15] Producers will be most protectionist when foreign films compete directly in a small home market for limited revenues. In contrast, distributors trading in

Figure 2.1 Commercial Filmmaking

Subsector	Production	Distribution	Exhibition
Activity	Making Movies	Selling Movies	Showing Movies
Ties to Foreign Markets	Weak and Indirect	Mixed	Strong and Direct
Trade Preferences	Protectionist	Mixed	Liberal

foreign films should have a more liberal outlook. This is especially true for those distributors who trade heavily in either foreign films in the home market or domestically made films in foreign markets.[16] Finally, exhibitors are likely to be the most liberal, particularly those who rely mainly on foreign films. Their preferences reflect a vulnerability to decreased revenues from any policy shift in the direction of greater protectionism. The relatively high asset specificity of theater-owning exhibitors also limits the extent to which they can press independently for their preferences, given the high costs of converting theater assets to other profitable activities and the difficulty of exiting.[17]

Any existing structural connections among these three subsectors realigns preferences and raises the industry's capacity to make effective demands when facing import pressures from powerful foreign competition. In the absence of perfectly competitive markets, firms and workers find themselves better positioned to intervene through the political process to secure their demands. Under oligopoly, when there are relatively few firms and an extensive network of linkages among producers, distributors, and exhibitors, the sector as a whole is better able to speak with one voice to state policy makers. This is even more evident under monopoly, with complete control of the sector by either state-owned enterprises or private firms, since there is the fullest development of an integrated, unified market structure and the greatest sectoral capacity to make demands. In this instance, film producers need not worry about distributors shunning their films; distributors are not concerned about exhibitors buying only popular foreign movies; and exhibitors are assured a steady stream of films to fill their schedules. The three subsectors can work cooperatively toward shared goals.

Alternatively, a highly competitive market structure reduces sectoral lobbying capacity significantly. The absence of ownership connections across subsectors and among the typically large number of firms generates serious collective action problems and raises the level of intraindustry conflict. Film producers find themselves at the mercy of distributors, and exhibitors are free to purchase whatever films they believe will attract the largest audiences. In some cases, sectoral and interfirm connections are limited by antitrust laws, which prohibit ownership across subsectors and break the monopoly power of dominant, strategically positioned firms. This occurred in the United States after the Paramount Decrees in 1948 forced the major producer–distributor combines to divest themselves of their exhibition holdings, prompted largely by exhibitors' complaints and the rising political power of small business.[18] A high level of domestic competition, then, mediates societal responses to international competition by undermining both the will and capacity of distinctive subgroups to push for protection.

According to the sectoral logic described here, when faced with intense import pressure, a less (or more) competitive market structure generates a more protectionist (or liberal) trade policy. Heavy foreign imports elicit fierce resistance from united industries. Divided industries either cannot or do not wish to resist imports. Changes in market structure, including shifts in state ownership, therefore explain the varying capacity of societal actors to promote their interests successfully.

Institutional Parity

In addition to affecting the flow of imports and exports (trade policy), globalization raises questions about whether the state can and should regulate the substantive content of national cultural production (cultural policy). State authorities responding to globalization are constrained and enabled by the existing institutional structure in the country. Institutions create and support bureaucracies, generate political constituencies, and affect the definition of state interests, thereby influencing how policy makers react to global pressures.[19]

In any given polity, the degree to which state cultural institutions are structurally equivalent to their economic counterparts shapes cultural policy decisively. Past choices about the relative power of the state in this domain affect the aims and aspirations of present state policy makers. While resolute leaders in new regimes or facing new political circumstances may alter the structural parity of state institutions, they cannot do so easily and without cost. This is especially so in the realm of national cultural production, putative threats to which can be manipulated—sometimes cynically—to damage political rivals. Changes in cultural policy therefore lag behind regime change and other shifts in the domestic political context, occurring only after institutional parity is transformed.

The most critical feature of state institutional arrangements for the film sector is the extent to which policy making in any given state accommodates the cultural significance of film. This significance is not universally recognized. In the United States, for example, the film industry is regulated as part of the entertainment business, and motion pictures are widely viewed as economic commodities—private goods—like all others. The most prominent state institutions that affect the industry's operation, such as the Department of Commerce, tend to support the pursuit of purely financial objectives, as opposed to furthering a well-defined set of national cultural goals. The National Endowment for the Arts notwithstanding, the United States has no federal Department of Culture or its equivalent, and state preferences reflect a clear laissez-faire orientation regarding filmmaking.[20]

Elsewhere in the world, in contrast, policy makers place filmmaking under the rubric of the "cultural industries." State regulation incorporates broader, noneconomic goals that include a conception of film as art, culture, and to some extent, a public good. In France, for example, state officials recognize and promote the identity-forming, educational, or ideological aspects of film more fully than their U.S. counterparts. Some of the key officials responsible for overseeing French policy reside in the Ministry of Culture. The institutional variation that distinguishes French and American approaches to film regulation is significant: It reveals whether regulatory bodies are responsive only to economic criteria, or if their "social purpose" is to promote cultural goals even when doing so entails economic costs.[21]

State regulation of the film industry is inevitably a mix of economic considerations and cultural concerns. While the economic domain is often privileged, the extent of this dominance depends on the structural relationship between state institutions in the two areas. In the case of a high level of institutional parity, economic considerations do not automatically trump cultural ones in policy making. State regulators may choose costly, economically inefficient policies because the individuals given the authority to make choices work for a Ministry of Culture and not a Ministry of Trade. More often than not, such an institutional configuration represents a legacy of the past, not the intended, conscious creation of contemporary state elites. Regardless of its origins, the degree of institutional parity between economic and cultural institutions is a major influence on the preferences of state decision makers. It accounts for the cultural dimension of policy responses to globalization pressures.[22]

This aspect of the argument generates a pair of interrelated claims. The greater the structural parity of the state's cultural and economic institutions, the less that policy makers regulate cultural industries using solely economic criteria, and the more that cultural policy is promotive in orientation. A promotive cultural policy entails active support for the industry through screen-time quotas, censorship regimes, and other interventions aimed at shaping national cultural production. The economic costs of a promotive cultural policy are borne, sometimes but not always willingly, by consumers and taxpayers, who are least able to organize to counter it and in whose name state institutions claim to speak.

Conversely, the less the structural parity of the state's cultural and economic institutions, the more that policy makers regulate cultural industries using solely economic criteria, and the more that cultural policy is laissez-faire in orientation. This implies that filmmaking is treated like all or most other industries. Under these conditions, the industry is promoted actively only when influential policy makers see an economic advantage to it (such as if a significant number of jobs are at stake), when the industry is considered strategically significant (perhaps for morale or propaganda purposes in wartime), or when the idea of protectionism is firmly entrenched in state *economic* policy-making institutions (as with import substituting development strategies). The economic costs of cultural production are borne by the industry, which orients its work toward the largest audiences possible in commercial cinema.[23]

Trade and Cultural Policy in Film

Trade and cultural policy are related state activities that always overlap, such as when cultural policies like cinema screen-time quotas have protectionist consequences for trade. Yet they are analytically distinguishable from each other, reflecting distinct political interests and ideas that usually are embedded in different institutional structures. Just as the more general rubrics of economic and social policy are not reducible to each other—despite their clear mutual influence—trade and cultural policy must be disaggregated to understand the larger issues at hand.[24]

Trade policy encompasses all official efforts to regulate foreign commercial exchange, including any state action that alters the price of imports or exports.[25] It varies along a single and continuous spectrum that ranges from protectionist to liberal. Policy in the film sector is expressed in the many forms noted in Table 2.3, with a state's general orientation toward the industry seen in their combined effects. Aggregating indicators avoids

Table 2.3 Trade Policy Indicators in the Film Sector

	More Protectionist	*More Liberal*
1. Tariff levels	Higher	Lower
2. Import quotas	Higher	Lower
3. Box office taxes on imports	Higher	Lower
4. Subsidies for domestic producers	Higher	Lower
5. Remittance restrictions	Higher	Lower
6. Import licenses	Present	Absent
7. Foreign exchange controls	Present	Absent
8. Intellectual property protection	Weak	Strong
9. Foreign direct investment	Not permitted	Permitted
10. Joint ownership	Not permitted	Permitted

the twin pitfalls of missing data and incommensurability, lending itself more readily to broad cross-national comparison.[26]

Cultural policy, in turn, includes any state effort to shape national identity by regulating cultural production.[27] Paralleling trade policy, it varies from promotive to laissez-faire, reflecting the extent of the state's commitment to developing and advancing a national cultural interest. Cultural policies include providing (or denying) state financial support for national cultural production, intervening in (or ignoring) its substantive content, and controlling (or disregarding) the availability of production. In the film sector, a promotive cultural policy may entail sustaining an otherwise inviable industry, influencing the content of domestically produced films through the censorship regime, or affecting the local availability of foreign and domestic movies via screen-time quotas. More laissez-faire policies would defer to private actors for financial investment in the industry and would permit market forces to determine the substance and availability of cultural production. State policy is generally promotive or laissez-faire, depending on the overall direction of the indicators present in Table 2.4.

State policies in the film sector may have surprising and sometimes perverse effects on the industries and actors they are intended to help. Under promotive cultural policies, for example, state subsidies yield a diffuse and subjective cultural gain to the public through the wider availability of films made by producers who otherwise would not be economically competitive. Benefits also may accrue to the foreign and domestic firms that take advantage of state support for what are essentially cultural goals. In this sense, purely rent-seeking firms may win subsidies under the guise of furthering cultural policy, and foreign producers may gain from quotas that have the effect of raising prices. The state extracts tariff revenues at the expense of foreign firms and tax revenues from fees assessed on admission tickets,

Table 2.4 Cultural Policy Indicators in the Film Sector

	More Promotive	*More Laissez-faire*
1. Screen-time quotas	Higher	Lower
2. Censorship regimes	Stronger	Weaker
3. State-financed production	Present	Absent
4. State-sponsored awards	Present	Absent
5. National film festivals	Present	Absent
6. Language requirements	Present	Absent
7. State film schools	Present	Absent

nominally intended to benefit local producers. The distribution of the costs and benefits of responses to global pressures depends on the exact instruments used to implement state policy, but the policy often has unintended consequences.

The Logic of Cross-Regional Comparison

Subsequent chapters demonstrate the argument's persuasiveness by analyzing the domestic and international dynamics of the film trade and comparing in detail the industries in Egypt and Mexico. A comparison of Egypt and Mexico is useful for this purpose, given their considerable international-level similarities, the nature of their domestic differences, and the different trajectories taken by their trade and cultural policies over the years. A cross-regional comparison also offers valuable analytical leverage that otherwise is not possible, since regionally dominant film producers are much more comparable to their counterparts in other areas of the world than to lesser producers in the same region. Comparing states that are similarly situated in the international film trade controls for alternative systemic explanations, while revealing how domestic factors shape policy outcomes.

Egypt and Mexico hold structurally and historically comparable positions in the world film trade. Film industries in both countries were established early this century; by the 1930s, they emerged as the leading regional exporters, respectively, to the Arabic- and Spanish-speaking markets of the Middle East and Latin America. Both regions integrated into global markets very early. By the mid-1940s, motion pictures were a significant source of export earnings and cultural influence for Egypt. At the same time, the Mexican film industry sustained production levels that dwarfed those of its closest rivals in Argentina and Brazil, with Mexican actors and directors winning critical acclaim internationally, as the industry's sizable export earnings grew. While in both instances their successes have since faded, Egypt and Mexico were regional leaders in the film trade throughout the twentieth century, each experiencing periods of severe crisis, heavy state intervention, decline, and eventual renewal.

The Egyptian and Mexican industries, moreover, have faced similar international pressures from their American competitors in the trade. The importance of the United States in this area cannot be overestimated easily. Despite the early achievements and linguistic advantages of local Egyptian and Mexican producers, the American industry never has lost its dominant position in their regional markets, maintaining a share of national screen time at first-run theaters that has ranged from 60 to 90 percent. These two

Table 2.5 Trade and Cultural Policies for the Egyptian and Mexican Film Industries

	1930s-40s	1950s-60s	1970s-90s
Egypt	Liberal-Laissez-faire	Protectionist-Promotive	Mixed-Laissez-faire
Mexico	Protectionist-Promotive	Mixed-Mixed	Liberal-Laissez-faire

midsized film-producing countries always have existed in the shadow of powerful international competitors, who in recent years have played an integral part in the globalization of the industry. While Mexico's closer geographic proximity to the United States has fostered ties between filmmakers in the two countries, the American presence in the Egyptian market has been sufficiently similar to justify the comparison.

Despite these similarities, there have been long-standing and puzzling differences in Egyptian and Mexican policy responses. Faced with similar pressures, film policies in the two countries nonetheless have taken very different trajectories over the past several decades. In the 1930s and 1940s, Egypt's film market was relatively more open than that of Mexico, where various types of protectionist measures were employed. The 1950s and 1960s saw dramatically growing Egyptian state involvement in the industry, and a partial diminishing of the Mexican state's role. Crises in both industries by the late 1960s led to opposite outcomes, with Mexican nationalization of filmmaking in the 1970s accompanied by a period of modest Egyptian liberalization. A full reversal from the past occurred by the 1980s and 1990s: significant liberalization of the Mexican industry and a withdrawal of the state, while filmmaking in Egypt remained heavily regulated, subsidized, and controlled by state authorities. Table 2.5 describes this variation.

What explains these different policy trajectories? The following chapters highlight crucial differences in market structures and state institutions, while ruling out leading alternative explanations. In periods when Egyptian filmmakers existed in competitive domestic markets, industry demands for protection from American imports were weak and liberal policies ensued. Changes in the institutional context, such as the expanded administrative authority of the Ministry of Culture and National Guidance in 1958, led to greater state regulation of filmmaking on noneconomic grounds and facilitated the growth of promotive cultural policies throughout the 1960s.

Similarly, Mexico's highly rigid and protectionist industry only gave way to a more open set of policies when economic crisis prompted a shift in market structure and state institutions, undercutting the protectionist coalition of filmmakers and their state allies. Global competition in the film industry was unrelenting, but state policies varied considerably, mirroring the precise nature of sectoral and institutional connections among key domestic actors.

Part II

States, Markets, and Competition

3

International Competition in Film

An extensive international trade emerged very early in the history of filmmaking, with the economic characteristics of the industry leading producers and distributors to seek expanded markets overseas. Competition in the film trade affected even the smallest commercial industries worldwide, provoking a range of state responses while highlighting the ambiguous and complex nature of film. In addition to providing vital historical context and revealing the tension between film commerce and culture, an understanding of international competition is a necessary first cut at explaining policy responses. Problematizing competitive pressures illuminates their actual effects on the industry and their analytical limits in accounting for state responses to globalization.

I begin by considering why film is bought and sold so extensively in international markets, elaborating on its most noteworthy characteristics as a tradable commodity. I then detail the unfolding of three major phases in film-trade history: an early period of U.S.-dominated oligopoly (1930s–1940s), a middle period of crisis and transition (1950s–1960s), and a late period of globalization (1970s–1990s). For each period, I recount key developments in the global and regional contexts, describing their effects and explaining their analytical significance. Globally, I link a range of conditions facing the American industry to changing levels of U.S. exports. Regionally, I focus on American sales to Egypt and Mexico, as well as Egyptian and Mexican exports to markets in the Middle East and Latin America. If international competitive pressures are as significant as the conventional wisdom suggests, they alone would explain important variations in trade and cultural policy. I show in the last section why this is not the case.

Trade in Cultural Commodities

A complex array of factors shape the trade in motion pictures and account for the direction and volume of film exports. These factors differentiate film from some tradable commodities while associating it with others, including an emerging subset of goods in the service sector of the global economy. They include the "nonrivalness" of film as a commodity; the need to reduce risk and uncertainty in the business; the expression of cultural power through filmmaking; and the effects of "language markets" on trade patterns.

Nonrivalness

Quintessential public goods include national defense, environmental protection, and urban infrastructure like sidewalks and roadways. They are provided by states acting ostensibly in the public interest, and their economic impact is felt primarily in the domestic context. In fact, some exports have the "nonrival" quality of public goods because their international provision does not diminish their domestic availability. These include exports with a cultural, ideational, or "knowledge"-based component, such as commercial forms of public culture, technological discoveries, educational models, and development paradigms.[1]

All motion pictures have the characteristic of nonrivalness because their consumption by one moviegoer does not affect their availability to others. Filmmaking, along with television and radio programming, is a quasi-public goods export by virtue of the strength of its nonrival economic characteristic. Regardless of the broader industrial or artistic circumstances under which any given film is made, production costs are always concentrated in the making of a single and unique master negative or original. From this negative, an unlimited number of prints or digital copies may be made at a small fraction of the original's production cost and distributed to exhibition markets. Hypothetically, such prints may be sold in an infinite number of markets worldwide without either affecting the price of the product to "rival" domestic consumers or diminishing the film's availability in home markets.[2]

An important implication of the nonrival aspect of film production is that it creates a compelling economic incentive for producers to export their films to as many markets as possible. With domestic markets unaffected by foreign sales, the opportunity cost of failing to distribute widely creates a powerful export imperative. Most of the alternative financial investments available to film distributors will yield lower returns than those to be obtained from the international distribution of a completed picture. Firms that have a capacity to sell abroad have an advantage over those that do not. This is not to claim that all films, all of the time, can be exported. Budgeting and thematic choices may limit the likelihood of international success and preclude the export option. Yet inherent in the relative distribution of costs is an incentive to distribute as widely as possible at home and abroad. Commercial producers never abstain intentionally from exporting a potentially profitable film.

In purely economic terms, filmmaking is therefore more comparable to the writing of new software or the manufacture of pharmaceuticals, for example, than to the production of automobiles. High initial expenditures for each new film "unit" are analogous to research and development costs, which are recouped by sales in as many markets as possible. Unlike the labor- or material-intensive production in automaking, the marginal cost of producing additional film units for sale does not come principally from labor or raw materials, which have already been sunk into production. A picture that costs thirty million dollars to make can be "sold" profitably to a foreign moviegoer for a dollar. It has very different dynamics in comparison to an automobile that costs fifteen thousand dollars to make but that must be sold to a foreign driver for perhaps twenty thousand dollars.

For this reason, only film distribution costs need to be factored into calculations of their export potential. Overseas marketing campaigns can be expensive, but the transport of films, even internationally, is not very costly in relation to potential revenues from exports. Such costs are particularly low because the ratio of weight to value for film is small in comparison to that of many other exports. In the film trade, such costs have been declining throughout the century, as transportation options have expanded to include new methods like airfreight.[3] Technological developments over the years have only enhanced the importance of the nonrival aspect of filmmaking, and world trade in this cultural commodity has increased accordingly.[4]

Risk and Uncertainty Reduction

If the nonrival aspect of filmmaking heightens the desirability of exports, another of its characteristics creates an even stronger incentive to trade in films: the industry is marked by exceptional financial risk and uncertainty. In terms of consumer demand, any commercial enterprise dealing in cultural or artistic production is subject to the particularly unpredictable and fickle nature of public taste. Predicting box office success is notoriously difficult.[5] Likewise, on the supply side, filmmaking does not lend itself to easy control of labor costs. As with all cultural production, scientific management techniques cannot determine with accuracy the nature and quantity of essential labor inputs.[6]

Filmmaking therefore is a risky business because human creativity defies automation while eliciting an uncertain public response. This risk is especially acute since investment decisions have to be made without reliable foreknowledge of returns. Each new film represents, in effect, the first and last stage of a production process that has the potential to meet with either great financial success or utter failure. In reality, the majority of commercial films yield net losses, at least nominally, with producers attempting to offset such failures over time by a small number of outstanding successes.[7]

As a result, industry strategies to reduce risk and make success more likely and predictable are a fundamental, overriding concern. They govern most other choices in the business. Such strategies have included the vertical integration of the industry and the development of "star systems" and genre conventions. Successful star systems, for example, create a loyal audience of fans who are likely to see any picture in which a given personality appears.[8] Genre categories allow new movies to be sold on the basis of tenuous connections to older successes. The effectiveness of such strategies lies in stabilizing demand, thereby bringing predictability to the flow of returns on investments. For its part, wide distribution enhances the likelihood that overall returns will be sizable, since any given picture may fail in some markets but do well in others. For relatively little additional cost, the potential benefits can be enormous. Risk reduction therefore requires that a film be exported whenever possible.

While filmmakers face strong incentives to distribute widely, doing so is feasible in the film sector because distributors can engage in price discrimination, selling their films at whatever price individual foreign markets will bear.[9] Most consumers in developing countries cannot afford to purchase expensive manufactures at uniform world-market prices; however, film export prices can be pegged to local economic conditions, which facilitates exports. The incremental revenues gained from small markets worldwide often contribute decisively to the profitability of a given film, even if smaller film producers in target markets have denounced such practices as "dumping," aimed at driving them out of business.[10]

Cultural Power

State power is a third factor that affects the film trade significantly. It is a truism to claim that powerful states in history have influenced both the shape of the world economic order and its major forms of cultural expression. Hegemonic states have maintained and expanded their influence through both commercial ties and overt political domination, and such power has translated into subtler kinds of influence, including in the cultural sphere. The film trade is no exception to this historical tendency, expressing the cultural power of leading international actors like the United States. As early as 1916, the U.S. industry achieved a degree of market dominance in filmmaking, holding a preponderance of power and resources that has translated since then into perhaps half of the world's first-run screen time. The U.S. position in world cinema has been one of the more pronounced and visible indicators of American power, sometimes spurring political rivals such as the Soviet Union to compete for influence by establishing a cinematic presence in major U.S. markets overseas.

The key to American cultural power is not simply the size of the U.S. domestic market, as critics often contend. U.S. producers and distributors have used their vast home market to further their trade position throughout the entire world, but their capacity to do so has rested on the broader foundations of American political and economic power. By no coincidence, the growth of American power in the early twentieth century was followed by its extraordinary rising influence in world culture, media, and the arts. In the film sector, this influence has a specific and traceable history that began with the ascendance in world commerce of modern American business enterprises and was abetted by concrete support for the industry by the U.S. government. The capacity of major American film distributors to establish hundreds of overseas branches was not unrelated to the political and economic power of the United States to underwrite such endeavors. The costs of establishing systems of foreign distribution were subsidized, assistance that most other governments could not offer.[11]

Even if on a more limited scale, regionally dominant states also have had the capacity to promote film exports and assert their cultural power in their neighborhoods. The Egyptian, Indian, and Mexican states all have supported, at one time or another, their exports to the Middle East, parts of Asia, and Latin America. When and where they have not done so, local industry has declined, as occurred with the Egyptian retreat from regional film exports in the 1970s, induced partly by its shifting position in Middle East politics. Where regionally dominant actors have challenged larger world powers, or found themselves competing directly with them, their limited influence has been most apparent, as in Mexico's forced withdrawal from postwar Latin American markets in the face of U.S. competition.

Even the most relentlessly commercial films are laden with semiotic content that may elicit political contestation. In contrast to otherwise comparable service-sector products such as computer software, the cultural power underlying film adds a politically charged factor to debates over trade policy, provoking both commercial and ideological resistance in local markets. The roots of the debate extend deep into the past century, though the American position has been unwavering: in exporting public culture for popular amusement, Paramount Pictures once operated under a provocative but telling motto—"We Entertain the World"—revealing both the global scope of its ambitions and the presumed ideological neutrality of its films.[12]

Language Markets

Industry observers have long noted the competitive edge held by those national film industries with large domestic audiences, especially those producing in languages spoken by substantial populations residing in foreign markets. The size of a given "language market" is therefore a uniquely important determinant of comparative advantage in the film sector: those industries serving the most speakers of a world language have the largest potential audiences for whom to make films without the costs and diminished appeal of subtitling or dubbing.[13] The vast size of the American domestic market is a noteworthy example, aided by the prominence of English as an international language. Producing for a large and comparatively wealthy domestic audience has enabled the American industry to make films with much higher production costs and then to export these high-value goods very successfully.[14]

Missing from this account, however, is a recognition of the extent to which seemingly natural language markets are conditioned by other, causally prior, political factors. Liberal assumptions of the naturalness of markets neglect consideration of the political conditions under which markets are constructed—or not constructed. Regarding film markets, the role of English as the world's most common second language is not unrelated to British imperialism or the more contemporary exercise of American power. France's substantial export markets in its former colonies are testimony to the significance of politically constructed and assiduously cultivated economic and cultural ties. The absence of export markets for historically prolific film industries such as Japan's has stemmed from a failure to create and maintain external connections of this kind. The strength of international markets must therefore be problematized in a manner seldom considered in conventional economic analysis of the industry.

An added but vital aspect of the international market is the range of purchasing power and the consumption habits of individuals in different contexts. Economies with individuals who can devote substantial resources to purchasing luxury goods like entertainment give a measurable advantage to the industries that serve them. Trade and consumption patterns reflect, for example, differences in per capita gross national product, with a high density of international economic transactions among the developed economies of the industrialized North. In countries like China and India, however, the pattern of economic development and heavy state intervention has precluded higher levels of consumer spending. Population alone clearly does not translate into large domestic markets. The politics underlying the purchasing power and consumption habits of individuals complement the role of language markets in world trade.

Trade Dynamics and Consequences

The evolution and consequences of international competition in film have reflected the imperatives of nonrivalness, risk reduction, cultural power, and language markets. The industry's gradual integration on a global scale proceeded steadily in the past century. Paralleling other sectors in the world economy, the most successful firms moved from nationally based filmmaking to more internationally oriented production and trade. For such firms, this process of internationalization culminated in the advent of a highly globalized industry by the 1970s, with strong interdependencies of markets, finance, and labor. For the remaining firms in the world, the opportunities and constraints they experienced were influenced substantially by developments in their more powerful rivals. An account of the principal developments in the film trade therefore is weighted heavily toward outlining the efforts and challenges facing the major firms, most of which were American in origin.[15] Table 3.1 summarizes the major trends, including changes in the industry and in the extent of its globalization.

Table 3.1 Globalization of the World Film Trade: Periods and Characteristics

Period	*Degree of Globalization*	*Industry Structure and Leading Technology*	*Description*
1930s-1940s	Low	Vertically integrated oligopoly in the United States; Fordist mass production.	Global expansion & distribution; Creation of national industries.
1950s-1960s	Moderate	Decentralization of production; Rise of television; High-cost technological innovations.	Increase in trade & dependence on foreign markets.
1970s-1990s	High	Full global integration; Flexible specialization; Digitization.	Multinational mergers; Global production, financing, & marketing.

Early Years: Industrial Rise and Consolidation

Filmmaking was recognized early on in the United States for its significant commercial potential, though not immediately as an export commodity. Within a few years of the invention of Thomas Edison's Kinetoscope in the early 1890s and Louis and Auguste Lumière's *cinématographe* in 1894, the first commercial projectors of moving pictures—movies—began to transform the phenomenon from a simple curiosity into a widely available and profitable form of mass entertainment and cultural expression.[16] Although French film companies like Pathé were the most active suppliers of the American market in the first decade of the twentieth century, World War I caused a precipitous decline in French production. The opportunity to gain a foothold in the European market was made possible by the oddly fortuitous calamity of the war, which destroyed much of Europe's productive capacity and provided an entrée to the nascent American industry.[17]

Initially, American producers were slower than their European rivals in pressing for exports for two reasons: the immense U.S. domestic market afforded more than enough outlet for early production; and perhaps more importantly, a series of intense struggles over patents and technology created an uncertainty that militated against making heavy overseas commitments.[18] Still, by 1918, American companies began to fill the void created by France's diminished production and oversaw a shift from London to New York as the film finance and distribution capital of the world. The growing power of the American industry in the silent era was reflected in the unparalleled worldwide popularity of United Artists' founders Charlie Chaplin, Mary Pickford, and Douglas Fairbanks. These first icons of American popular culture created the earliest international pressure on local culture producers by integrating their smaller markets into the domain of American exports. They also raised acute concerns on the part of business-minded European elites, who repeated approvingly a phrase attributed to England's Prince Edward in a 1923 speech: "Trade follows the film."[19]

The rapid expansion of overseas activities by the American industry was aided greatly by the formation of a trade association, the Motion Picture Producers and Distributors of America (MPPDA), established by a group of industry leaders in March 1922. Created in accordance with the Webb-Pomerene Export Trade Act of 1918, the MPPDA was exempt from the antitrust provisions of U.S. law, such as the Sherman Act of 1890.[20] Embodying newly influential corporatist thinking about business-state relations, the Hays Office—as it came to be called for its tireless first chairman, Will Hays—sought to facilitate cooperation among the major U.S. firms on matters of domestic commerce, censorship, and foreign trade.[21] The MPPDA opened dozens of offices worldwide and created a Foreign

Department to act as the trade representative of the largest companies in negotiations with foreign governments. It attempted to counter the first trade barriers to Hollywood exports, when Germany instituted restrictive measures in 1927, eventually earning the nickname "the little State Department" for its extensive diplomatic activities.[22]

Perhaps just as significantly in the long term, American filmmakers in those early, opportune years did a great deal to define the medium of commercial film in the eyes of world audiences, establishing cinematic tastes for filmmaking in a linear, continuous, "classical" narrative style and creating expectations that only large production budgets could fulfill.[23] While other viable modes of film production were not foreclosed, by the 1920s, most non-American and noncommercial filmmaking was viewed as an alternative to the Hollywood norm. As national film industries were established elsewhere in subsequent decades, they nearly always sought to emulate the Hollywood style and aesthetic, as well as its organizational pattern, studio production mode, distribution methods, and star system.[24] These industries certainly developed nationally specific forms of artistic expression, but they still represented variations on a common theme.[25] Thus, American influence—both substantive and organizational—proved to be powerfully path dependent, as early success bred later success, and the development of filmmaking throughout the world was shaped by the partly random events that accompanied the birth and growth of the industry in the United States.[26]

1930s–1940s: Oligopoly and Global Expansion

A series of corporate mergers in the 1920s and 1930s led to the rise and consolidation of an oligopoly in American filmmaking, as the number of important firms competing with each other declined from several dozen to only five "majors" by 1930.[27] This was the Golden Age of Hollywood, or the studio era. Most significantly, these firms were vertically integrated through their involvement in all aspects of the film industry, from financing to production to distribution to exhibition. Vertical integration had implications for the way that films were made and marketed, both domestically and abroad, ensuring a steady flow of revenues through a conservative, systematic, fully rationalized approach to filmmaking based on careful audience research and the use of established stars and genre conventions. Anticompetitive practices, such as "block-booking" and producer-exhibitor combines, assured screen time for all studio pictures, while foreign and independently produced films were effectively excluded by their lack of connection to the larger industrial structure.[28]

The invention in the late 1920s of a reliable system of sound recording for film had accelerated industrial consolidation in the United States and was pivotal in furthering the American trade position internationally. Sound recording required much more extensive capital investment than silent films, reducing the number of producers available and contributing to the economic imperative to increase returns on investments by expanding film markets. Initially, it was a source of grave concern for Hollywood because sound filmmaking introduced the nationally specific element of the spoken language into the production process. But in short order, the development of sound gave American firms in the studio era two new trade advantages. First, unlike in capital-scarce interwar Europe, the availability of investment capital in the United States enabled American filmmakers to convert to sound rapidly, thereby enhancing the marketability of their product. Second, and perhaps more significantly, the size of the American market enabled producers to cover at home the costs of sound conversion, reinvest their returns, and expand exports abroad.[29]

Many European filmmakers found their trade positions undermined by a double dilemma in the early 1930s. On the one hand, their national markets were too small to earn enough revenues to convert quickly to sound and remain competitive with the high volume of American imports. But even if they succeeded in doing so, the change threatened them anyway because their languages were not spoken widely enough in international markets to sustain the more expensive production process. Some producers, moreover, only had ties to weak distribution apparatuses that were insufficiently developed to sell abroad effectively, even if marketable films could be produced. Industries organized on an artisanal basis simply could not compete globally with the modern, integrated business enterprises then hailing from New York. Locally made films found audiences in many places only by virtue of their novelty to those who wanted to hear their national languages emanating from the screen.[30]

Outside of Europe, the development of sound recording had implications for the emergence of new national film industries, especially in postcolonial states. Sound added a dimension to the medium that nationalist filmmakers could deploy to promote national identities, and that local entrepreneurs were well positioned to take advantage of by virtue of their knowledge of the cultural specificities inherent in language. Since sound filmmaking was relatively expensive, it sometimes invited state participation in investment. But with or without official support, new film industries appeared in European colonial holdings and postcolonial states, modeled in many cases on the American commercial pattern. Colonial powers like Britain and France were protective of their export positions in these overseas territories, mindful of their vulnerability to the American industry. Despite the economic interests at stake, however, their paramount concern early on was for the film medium as a potentially subversive cultural influence, and both colonial powers instituted stringent censorship regimes. For many years, the French prohibited native West Africans from attending the cinema. British authorities worried incessantly about the mass psychological effects of films depicting Britons engaged in untoward behavior or in weak positions vis-à-vis their adversaries.[31] Without the burden of direct imperial concerns, Hollywood was free to focus on building its own tightly controlled empire in the United States and abroad.

In Egypt, on the eve of the sound revolution in 1926, the American Trade Commissioner in the cosmopolitan port city of Alexandria reported to Washington that half the motion pictures shown in the country were American, with U.S. films experiencing no trade discrimination or political interference whatsoever.[32] Movies were rented to cinemas by the meter for a fixed price, with the Alexandria market covering two-thirds of local distributors' costs and Cairo responsible for the remaining third.[33] As a whole, Egypt at the time was a small but expanding market, second in the continent only to British South Africa.[34] In fact, the popularity of Hollywood film over European exports was just beginning a steady rise to reflect its growing worldwide dominance (see Figure 3.1).[35]

In conjunction with this rise, American cinema eventually would be drawn into the turbulence of local politics in the coming decades: twenty years to the day after the American Trade Commissioner's report, the Motion Picture Association of America forwarded to the State Department information regarding a grenade attack on the Cinema Miami in downtown Cairo by anti-British activists.[36] More commonly, pictures like MGM's *Song of Revolt* suffered at the hands of state censors for dealing with sensitive political themes deemed unsuitable for Egyptian audiences.[37]

Egypt's domestic film production dates to 1923, when German-trained filmmaker Muhammed Bayoumi returned to Cairo to shoot his first fiction film, *Fi Ard Tutankhamun*.[38] Bayoumi also made a short film, *al-Bashkateb*, at about the same time, and both pictures preceded what used to be considered the first Egyptian feature: *Leila*, a 1927 silent film starring a well-known stage actress, Aziza Amir. Sound production followed a few years later in 1932, with Muhammad Karim's *Awlad al-Dhawat*, written by and starring the wealthy, theatrically trained Yusuf Wahbi. Leila, in fact, was the first entirely Egyptian-financed film, since most of the earlier filmmaking was done by resident foreign nationals, whose ties to Egypt gave them somewhat ambiguous identities.[39] By the early 1930s, Hollywood already had come to dominate the local market, with American imports comprising 76 percent of all films screened in 1936, followed by France's 10 percent.[40] The country's geographic position at the crossroads of Africa

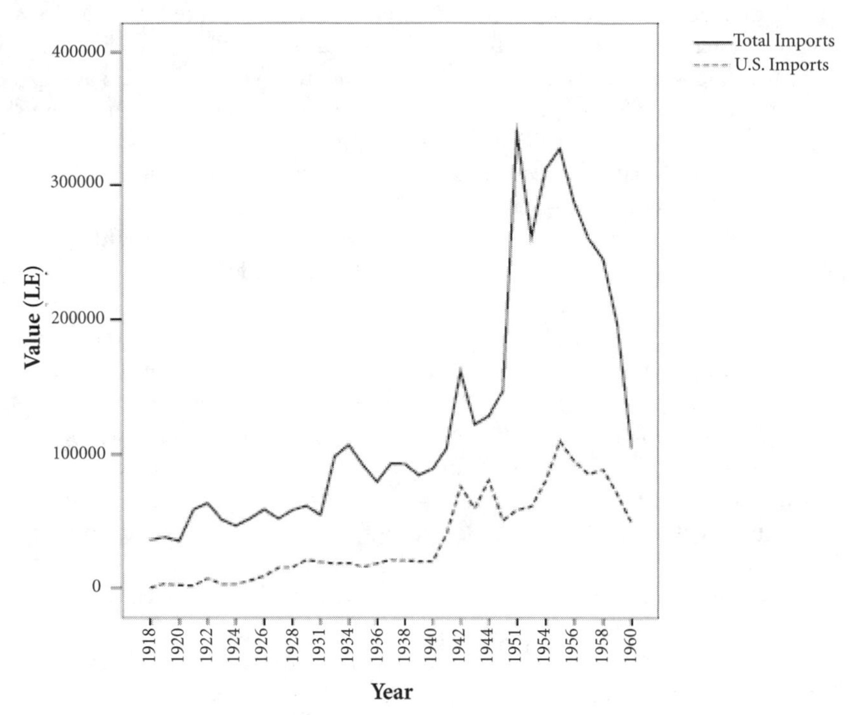

Figure 3.1 Egypt: Film Imports, 1918–60
Source: Compiled from the Annual Statement of the Foreign Trade 1918–60

and Asia magnified its significance in the film trade, since American distributors shipped films from the United States to Europe and then on to other markets via branch offices in Egypt. Universal Pictures, for example, maintained its Near East regional headquarters in Alexandria, which served as a transit point for film shipments to markets throughout the Middle East, Africa, and beyond.[41] American competition, then, was a defining characteristic of the Egyptian market from its earliest days.

Domestically produced Egyptian exports did not develop in earnest until after 1935, when prominent industrialist Mohammed Tal'at Harb built Studio Misr, the first modern film studio in the Middle East and Africa.[42] Egyptian production and export of feature films grew dramatically in the 1930s and 1940s, with Cairo emerging as a self-styled "Hollywood of the Arab East," quickly overshadowing Alexandria as the industry's center of gravity (see Figures 3.2 and 3.3).[43]

Entrepreneurs, some of whom were resident foreign nationals, constructed four more major studios by 1947, and the number of independent production companies quintupled from twenty-four to 120 by 1950.[44]

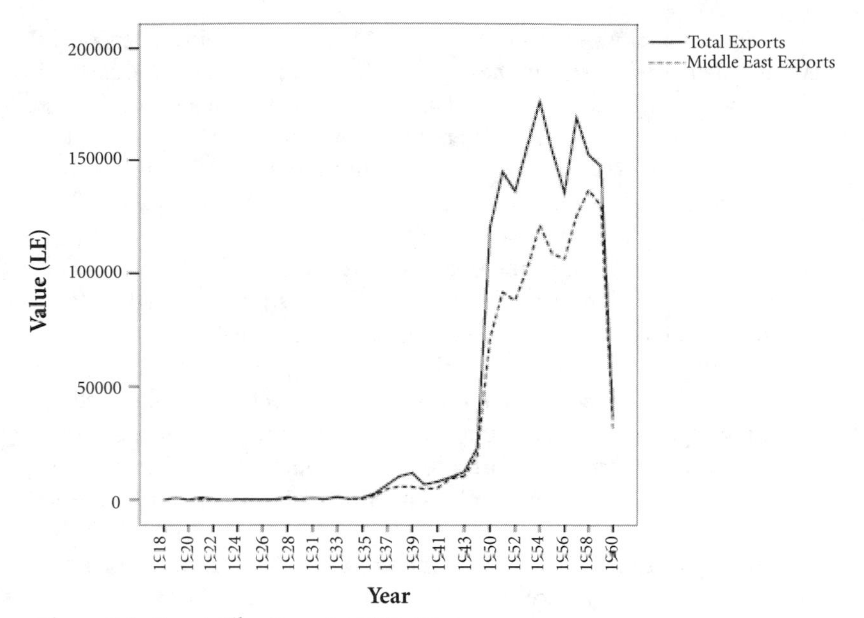

Figure 3.2 Egypt: Film Exports, 1918–60

Source: Compiled from the Annual Statement of the Foreign Trade 1918–60

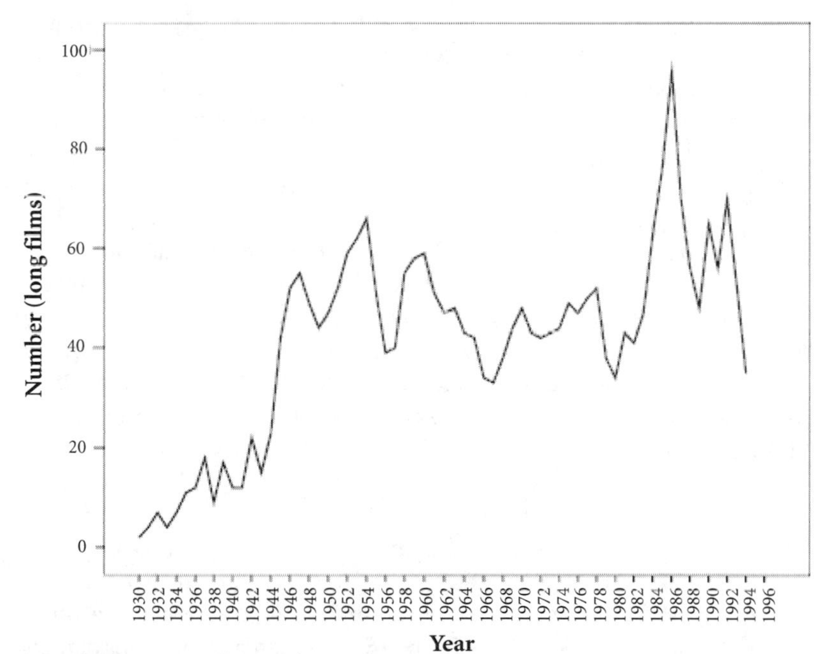

Figure 3.3 Egypt: Film Production

Source: Compiled from Federation of Egyptian Industries; Wassef; Ministry of Culture

Investment capital was most readily available during and just after World War II, when investors sought the quick profits of expanded production and diminished foreign competition. This was the Egyptian industry's first Golden Age, at least financially, when moviegoing became the most popular form of urban entertainment in Egypt and much the rest of the region.[45]

The profitability of film investments in this period had a crucial effect on the rapid commercialization of Egyptian filmmaking. The entrepreneurs drawn to the filmmaking business placed a very heavy emphasis on expanding production and extending foreign distribution, rather than improving quality.[46] Production companies rented out studios simply to make single films and distribute them as widely as possible, either by hiring a distribution agent to handle specific foreign markets or by selling limited foreign distribution rights.[47] Risk-minimizing strategies predominated in business decisions, mostly in the development of a rigid star system in which an exceedingly small number of actors appeared in multiple films. As a result, Egypt's actors and singers gained great popularity in neighboring Arab states and established a regional presence that echoed American global dominance, even if on a much smaller scale.[48]

Elsewhere in the world, American exports expanded throughout the interwar period. Mexico's close geographic proximity to the United States and its strategic position in relation to Latin American markets encouraged a greater U.S. presence in the 1920s, including the opening of distribution outlets for Universal, Paramount, United Artists, and 20th Century-Fox.[49] This presence was possible only after an important early episode in motion picture diplomacy, when the Hays Office defused a crisis in U.S.–Mexican film relations in 1922. The consistently negative portrayal of Mexicans in American pictures had led to a total ban on their import, but by dispatching a representative for nearly three months of negotiations, Hays secured an agreement to lift the ban in exchange for promises of greater sensitivity to Mexican national sentiment. This encounter appears to have led Hays to broader conclusions about the importance of avoiding offense to local sensibilities, and it presaged the MPPDA's effectiveness in direct diplomacy with foreign governments.[50] By 1925, Mexico had become Hollywood's eighth largest market by volume, accounting for two per cent of its foreign sales.[51]

Throughout the 1930s, the consolidated American major firms experimented with varied strategies for retaining their positions in the large and lucrative foreign markets of Latin America.[52] Concerned that English-dialogue sound films would not be exportable to this and many other regions, most of the majors began filming Spanish-language versions of their pictures, drawing on actors and directors from Spain and Latin America to make 113 of them by 1938.[53] To do so, Paramount refurbished in 1930 an entire studio complex at Joinville, just outside of Paris, devoting it to making, in assembly-line fashion, as many as fourteen different foreign language versions of its productions.[54] These films tended to fail commercially, however, since Latin American audiences preferred seeing subtitled films with popular Hollywood stars rather than the odd, eclectic mix of actors and accents from all over the Spanish-speaking world.[55] The U.S. industry's "Hispanic" films of the 1930s nonetheless were significant in providing technical training and stylistic socialization for a number of Mexican filmmakers who would later become prominent in their home industries.[56]

World War II did not limit American access to the Mexican market, with U.S. producers managing to maintain an overwhelming market share of 85–90 percent. Unlike many European states that were cut off from U.S. sales during the war, Mexico stayed open to exports, and even before the war's outbreak, antifascist sentiment on the part of organized labor and state censors from the Department of the Interior weakened Germany's trade position in film.[57] Significantly, however, the war-related themes that rose to prominence in American pictures of this period had less resonance, and therefore marketability, throughout Latin America.[58] As a consequence, local Mexican producers found an opportunity to expand both domestically and abroad, and they did so at the expense of declining European production and reduced American distribution.

Just before the war, a resurgence in domestic production had begun with the international success of a single film, Fernando de Fuentes's *Allá en el Rancho Grande* (1936), which spawned an entire genre and convinced investors that Mexican films could be extremely profitable in Latin American export markets.[59] As in Egypt, Mexican filmmakers benefited directly and indirectly from the war, since the irregularity of American film shipments to Latin America provided them with export opportunities, just as several sectors of the economy grew to meet rising wartime demand. Growth in the film sector was possible in large part because the United States favored its pro-Allied neighbor over neutral Argentina. Aware of the propaganda power of the motion picture, the State Department's newly formed Office of the Coordinator for Inter-American Affairs worked indefatigably under Nelson Rockefeller to foster cooperation between the Mexican and American industries in technical, financial and personnel matters.[60] Denying Argentina access to raw film stock, it assured Mexico the necessary materials, equipment, and markets to flourish.[61] Thus, the war years were instrumental in consolidating Mexico's position as the leading Latin American film producer. In 1949, Mexico was the first

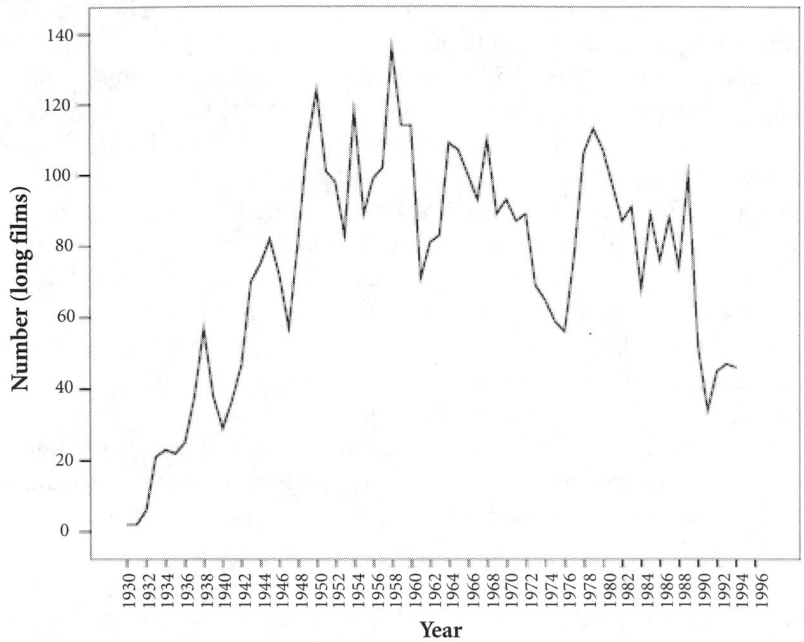

Figure 3.4 Mexico: Film Production
Source: Paranagua; Garcia Riera, 1986 and 1992

Spanish-language industry in the world to break the 100-film production barrier with a record 108 feature films (see Figure 3.4).[62]

1950s–1960s: Crisis and Conglomeration

American access to foreign markets assumed a new urgency in the postwar era. In May 1948, the U.S. Supreme Court ruled on the landmark *Paramount* antitrust case, mandating the end of vertical integration and the beginning of a new era of greater domestic competition.[63] This decision forced the majors in the subsequent five years to divest themselves of their exhibition branches and focus primarily on the making and distribution of movies, with the latter proving in the long run to be the strategic linchpin to the industry.[64] *Paramount* came at an inauspicious moment immediately after World War II, when film audiences were declining dramatically in the United States for several reasons: a postwar baby boom led more couples to stay at home with their children; rising educational opportunities for returning servicemen lessened the time available for moviegoing; and a spending spree on newly available consumer durables reduced the disposable income that could be spent on the cinema. These developments were accompanied by the precipitous rise of television in the early 1950s, all of which combined to lower levels of weekly movie attendance by about 50 percent by 1957.[65] The bleak domestic outlook was matched by a certain ambivalence among filmmakers about postwar European reconstruction, as the rebuilding of the continent promised both a renewal of consumer spending and a return of other national producers to challenge the U.S. industry's global hegemony.

Hollywood's response to this crisis had important consequences for film production and trade throughout the world. In addition to pursuing technological innovations like color cinematography and gimmicks like 3D viewing—essentially, strategies of product differentiation from television—the postwar American industry began a concerted effort to expand its revenues in foreign markets, partly in Europe but even more so in Asia, Latin America, and the Middle East.[66] This entailed a return to markets lost during the war and an expansion into underdeveloped areas in the subsequent two decades. While the global postwar dollar shortage did limit direct remittances of foreign revenues, the film trade was unlike those industries in which the strength of the dollar limited exports, since the use of price discrimination enabled American distributors to charge only what local markets could bear. Consequently, the percentage of total industry revenues coming from foreign markets rose from about a third in the 1930s to more than half by 1960, making the film sector one of the most fully internationalized in the world economy.[67] The internationalization of U.S. markets implied a substantial degree of dependence for the American industry, because without foreign earnings the industry simply could not have sustained the rising production costs that marked this period.

Beginning in the 1950s, moreover, the nature of American involvement in the international film trade began to change. Hollywood's postwar expansion intensified the pressure on foreign industries to maintain their market shares in the face of rising American competition, and some states responded with protectionist measures like remittance restrictions and currency exchange controls. Particularly in Europe, where investment capital was needed for major reconstruction projects, protectionism often left American firms with their earnings locked in frozen accounts, inducing them to find alternative ways of putting such resources to work. As a consequence, the American majors began to invest more directly in foreign productions to facilitate the repatriation of profits and to avoid other protectionist restrictions. In so doing, American-financed coproduction with

European filmmakers eased the burden of both the capital scarcity facing the latter and the newly stringent restrictions limiting the former.[68]

These developments had unexpected implications for the accessibility of the U.S. market to foreign producers. Long before, beginning with the first instance of film-trade protectionism directed against the United States in the early 1920s, American firms had based their call for free trade in film on the claim that no trade barriers were in place in the United States. Strictly speaking, this was true, but in fact the U.S. market remained effectively closed to non-U.S. producers due to the vertical integration of the industry and the reluctance of American distributors and exhibitors to deal with foreign films. With the breakup of vertical integration in the late 1940s and the subsequent rise of coproduction, however, U.S. distributors began to carry those foreign films that had partial American involvement. Distributors also attempted in the early 1960s to compensate for the weakness and instability of domestic demand by acquiring foreign exhibition outlets as guaranteed venues for their products in markets overseas. American exhibitors, for their part, no longer were constrained by their connections to a larger industrial network, and as a result, foreign films gained much greater access to the American market.

In the 1960s, integrative trends continued in the form not only of U.S. investment in European filmmaking, but in the building of studios and the undertaking of large-scale production activities overseas. These so-called runaway productions benefited from the increasingly overvalued dollar and relatively low foreign labor costs, even gaining access to generous state subsidies intended to support national filmmakers.[69] Much of this activity occurred in Europe, but industries in Africa, Latin America, and the Middle East were not unaffected by the shift. As American firms consolidated their dominance of distribution networks in the developing world, local filmmakers sometimes had to rely on American-based multinationals to distribute their work, both regionally and sometimes even domestically. Film exporters like Mexico, for example, lost their Latin American distribution networks to U.S. firms, which were better financed and had access to much larger film libraries that could provide exhibitors with a steady supply of popular movies.

Egypt was an important site of both contestation and collaboration in the film trade. Equipped with recently improved portable cameras and magnetic sound recording tape, American companies were drawn to the box office rewards of Egypt's "authentic" locations, the financial benefits of lower production costs, and any available opportunities to put frozen foreign earnings to work.[70] In the early 1950s, two examples of this were the on-location filming by Howard Hawks of Warner Brothers' *The Land of the Pharaohs* in 1953, as well as the shooting by Cecil B. DeMille of Paramount Studios' Academy Award-winning epic, *The Ten Commandments*, in 1954. Both were large-scale, expensive productions, the first using a popular new wide-screen process called CinemaScope and the second employing nearly 200,000 Egyptian artisans, technicians, laborers, and extras. They differed in their critical and box office successes, but both involved enough spending in Egypt to give Warner Brothers and Paramount temporary respite from their remittance problems.[71]

Despite the cooperative partnership implied by the latter productions, competition from American exports to Egypt intensified in the 1950s, even while Egypt's own exports gained momentum and expanded. U.S. producers hoped, in Egypt as elsewhere, to recapture the market segment lost in wartime, thereby compensating for declining domestic sales and steadily rising production costs associated with new technologies.[72] The American share of Egyptian imports peaked at 97 percent in 1957—307 of the 314 foreign films screened in the country—after the virtual elimination of French and British exports during the Suez crisis and their drastic reduction for several years thereafter (see Figure 3.5).[73]

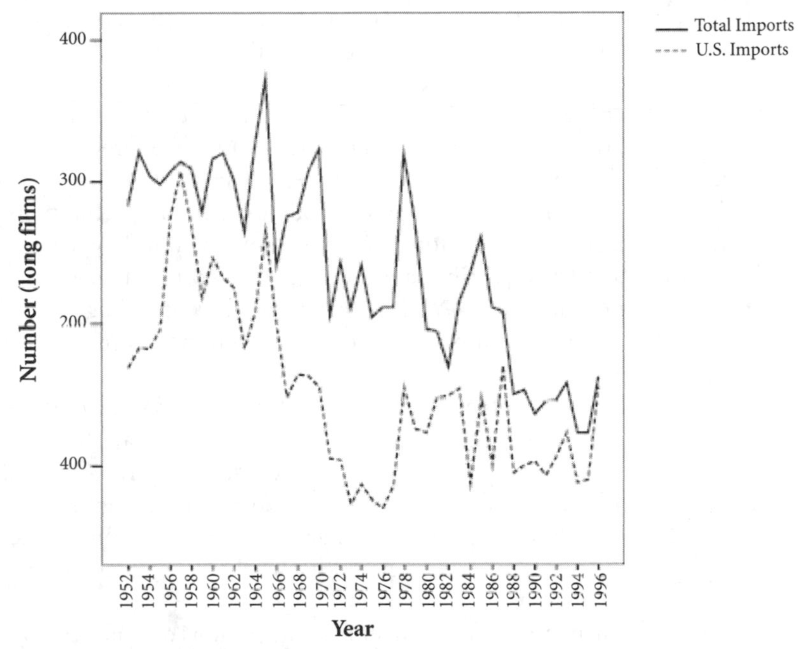

Figure 3.5 Egypt: Film Imports, 1952–96
Source: Paranagua; Garcia Riera, 1986 and 1992

From then on, however, U.S. imports began a two-decade slide downward, with American distributors closing or consolidating their Cairo offices in the 1960s to cut costs, maintaining distribution through local agents or their own representatives stationed elsewhere.

The postwar diplomatic record chronicles the efforts of American producers to defend their business interests by soliciting U.S. Embassy assistance, just as it tells the story of attempts by Egyptian producers and distributors to enlist government support in the face of foreign competition. Sometimes American film companies simply found themselves caught up in regional acrimony: Israel declared Paramount's *Samson and Delilah* to be anti-Jewish propaganda and banned it, at the same time that Egypt delayed the same film's release on the claim that it was pro-Zionist.[74] Generally, though, two international issues dominated the American government's film-related concerns in Egypt throughout the 1950s and 1960s: remittance restrictions on U.S. companies and broader cold war rivalry with the Soviet Union, the latter producing a more overtly political logic in Embassy support for American film companies. Embassy officers cooperated with Hollywood representatives, even if the written record contains evidence that officials took a broader view of U.S. interests and sometimes clashed with the purely business-oriented American film companies.[75] In the final analysis, American government and business agendas overlapped sufficiently to provide a substantial net boost to the U.S. competitive effort.

Egypt certainly was not alone in the world in prioritizing its domestic allocation of hard currency. But of all the concerns facing American distributors in Egypt in the 1950s, remittance problems were foremost, since an inability to repatriate profits deprived them of a return on investments already made and undermined any rationale for developing the market further.[76] Remittance restrictions originated in postwar Egypt's withdrawal from the British sterling area when the Anglo-Egyptian Financial Agreement expired in July 1947.[77] Egyptian uncertainty regarding the availability of dollars to cover imports led to a freeze and then simply a reduction in the amount of hard currency that American distributors could return to parent companies each month. Permitted remittance levels declined from 80 percent during the war to 50 percent in May 1947,[78] and finally to 35 percent under an agreement between the MPAA and the Egyptian government in January 1948.[79] The shortage of foreign exchange strained the Egyptian government's willingness to permit the remittance of even 35 percent of monthly earnings, at least until the global dollar shortage ended a decade later.

U.S. government assistance to the foreign operations of American businesses like filmmaking had roots early in the century, and postwar official support was driven initially by straightforward economic goals that differed from European concerns for political interests and influence in the world. The growth of cold war tensions, however, broadened and politicized official American backing for its overseas businesses in Egypt and other strategically significant locales. Evidence in the diplomatic record shows that the militarization and globalization of the containment doctrine, articulated by cold war planning instruments like NSC 68, had a cultural counterpart of sorts, manifested partly by U.S. efforts to promote American films at the expense of Soviet ones worldwide.[80] In the uncompromising atmosphere of the early cold war, public rhetoric about motion pictures being simply entertainment was disingenuous at best. Policy makers saw nothing as completely apolitical in a world where the Soviets invented a cartoon character of their own—a porcupine—to counter the worrisome international popularity of Walt Disney's colorful creation, Mickey Mouse.[81]

American film representatives in Egypt must have been aware of these broader developments, and they may have attempted to benefit from them by provoking the embassy with dramatic claims of Soviet market penetration.[82] In fact, the Soviet Union did increase its presence in the Egyptian film market, seeking indirect control over theaters by purchasing and leasing them through local intermediaries in the mid-1950s.[83] Within a few years, the Soviets had captured 10–15 percent of the Egyptian market, benefiting—though not as much as their American rivals—from the post-Suez diplomatic falling out, and taking advantage of Egypt's subsequent tilt to the left. An indication of the extent to which Hollywood became enmeshed in the larger political conflict of the day was the publication in 1957 of *Safir Amrika bil-alwan al-tabi'iyya* (The Ambassador of America in Natural Colors), a wide-ranging critique of American films and Hollywood's role in the world, written by left-leaning film director, Kamil el-Tilmissani.[84]

Despite the heightened international competition of the early postwar years, Egyptian film exports continued to expand abroad, largely sustained by the efforts of war profiteers to find fruitful investment opportunities for hundreds of millions of pounds in newly acquired wealth.[85] Still handled by individual sales agents, the majority of these exports went to markets in the Middle East and North Africa, though Egyptian films in the 1950s found their way regularly to such disparate and far-flung places as Venezuela, Hong Kong, Madagascar, Denmark, and Indonesia.[86] Closer to home, the high-water mark of Egyptian exports to the Middle East came in 1954, when Egypt shipped abroad over £E 176,000 in film, most of which went to Lebanon, Iraq, Jordan, Libya, Aden, and Syria.[87]

Just as Egypt's regional political influence grew, under the charismatic leadership of Gamal Abd el-Nasser, the Egyptian film had no true rivals in the Middle East and North Africa early in this period, though American pictures continued to win the lion's share of receipts in elite-oriented theaters. The official trade figures greatly underestimate the economic significance of exports by recording only nominal values for the physical reels of celluloid shipped abroad, unable to register their eventual returns from exhibition over a multi-year period. Using this crude method, film ranked as the country's twenty-second most valuable export in 1953, out of several dozen commodities.[88] In fact, if one observer is even remotely accurate in listing gross industry revenues at £E 2.7 million that year, the common assertion that film was Egypt's second ranking source of export revenue, after cotton, may not have been completely exaggerated.[89]

These successes notwithstanding, trade competition in the regional market eventually reduced Egypt's foreign revenues and was instrumental in precipitating its long-term decline. A sporadic but clear downward trend in exports began halfway through the 1950s, delayed only briefly by a boom period late in the decade that film historian Samir Farid has dubbed, "the second golden age for Egyptian cinema."[90] Foreign movies inundated the region, as the novelty of American color films and "spectaculars" combined with a rejuvenated European cinema led by the French New Wave. Egyptian producers and distributors faced competition unlike anything they had seen before. With the economic uncertainty that accompanied the Socialist Decrees of July 1961 and the partial nationalization of the industry that year, both production and exports trailed off considerably for most of the rest of the 1960s (see Figure 3.6).

Even if opinions vary on the effects of fuller state involvement in production after 1963, clearly it meant greater thematic didacticism that was not always well received by audiences accustomed to seeing film as simple entertainment.

Exports also were subject to the vagaries of regional politics, since inter-Arab rivalry affected the availability of markets for Egyptian pictures. While Iraq had become one of Egypt's most important foreign markets by the mid-1950s, political rivalry in 1960 appears to have led Baghdad to ban Egyptian movies, at least temporarily.[91] In the polarized regional climate of the mid-1960s, the film industry joined other politically engaged mass media, such as radio's *Voice of the Arabs*, causing regimes in Syria, Lebanon, Saudi Arabia, and some of the other Gulf states to slash imports from Cairo.[92] A few other Arabic-language producers like Lebanon began making a handful of films on their own, though the greatest beneficiaries of inter-Arab conflict were the American companies and affiliated local distributors and exhibitors that filled the void created by Egypt's departure.

General trade dynamics were remarkably similar in Mexico, on the other side of the Atlantic. Complementing the growing use of location shooting

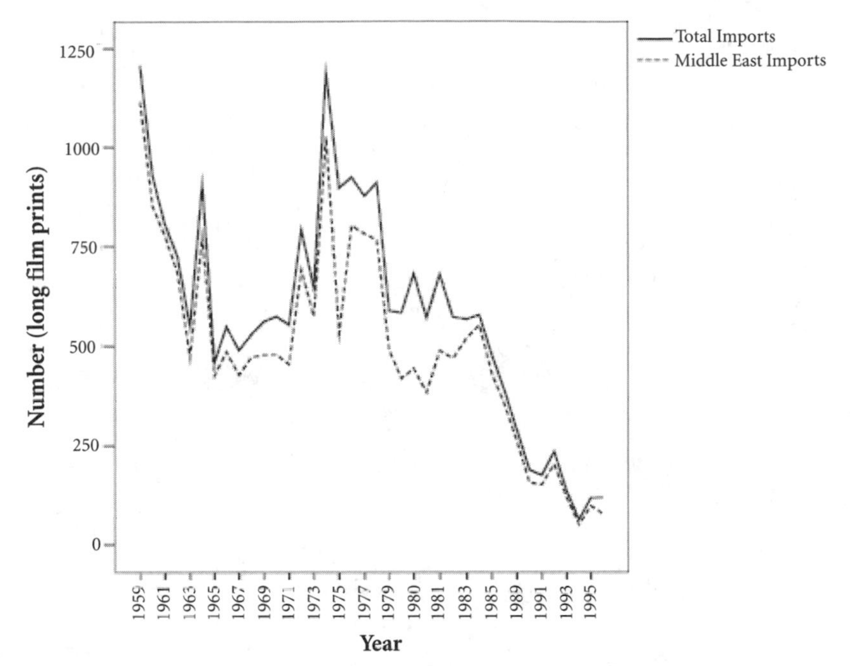

Figure 3.6 Egypt: Film Exports, 1959–96
Source: Compiled from CAPMAS and Ministry of Culture figures

worldwide, American companies like Warner Brothers and RKO initiated modest levels of production in Mexican studios in the 1950s.[93] The peso devaluation in 1954 and the availability of skilled labor and suitable local facilities enticed U.S. producers to make seven films in Mexico that year, followed by eight more in 1955, and a larger number of westerns in the north-central state of Durango beginning in 1958.[94] This trend differed from filming on location, since facilities and natural settings were rented out to make pictures that were thematically unrelated to Mexico as a place.[95] With the demise of the Hollywood studio system by 1960 and the growing prominence of independent production companies, Mexico's less costly working conditions offered an inviting alternative to filmmaking in the United States.

As they did in Egypt and the Middle East in this period, moreover, the American majors in Mexico attempted to regain their market share, which had declined somewhat by the end of the war. A resurgence of U.S. exports in the 1960s was made easier by a growing crisis in the local industry, as film quality deteriorated in increasingly formulaic productions (see Figure 3.7).

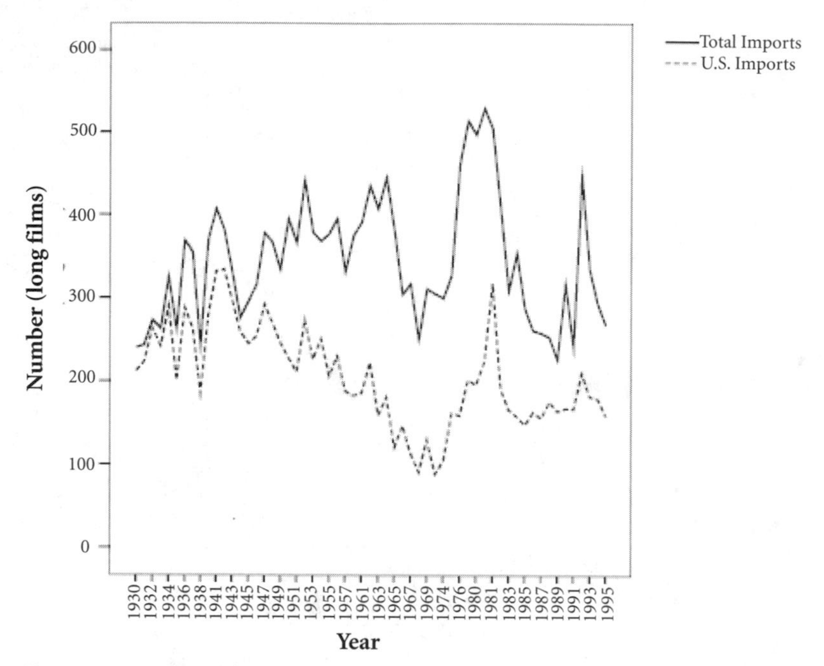

Figure 3.7 Mexico: Film Imports, 1930–94
Source: Compiled from Heuer; Garcia Riera; Amador; UNESCO

Such a crisis may have been inevitable in light of the extent to which many companies originally were drawn into the business by the exceptionally profitable opportunities of the wartime era. In the war's aftermath, these companies turned to making quick, low-risk, cheap pictures with no export potential and limited domestic appeal. Production levels remained high at first, but in 1950s Mexico, a growing share of the urban middle class grew disaffected and began to abandon national productions for Hollywood and other foreign fare.[96]

Not only did Mexicans themselves begin to lose interest in domestically produced films, but the crisis extended immediately to the industry's exports to Latin America. While Mexican films were acquiring a reputation for tawdriness, other major regional producers, such as Argentina and Brazil, were improving and augmenting their own production and, along with Spain, regaining a more substantial place in the Latin American market.[97] Regional instability and periodic political upheaval also took their toll, such as with the Cuban revolution's elimination of that market in 1958.[98] From the mid-1950s to the mid-1960s, moreover, the substantial devaluation of most Latin American currencies in relation to the Mexican peso—most notably those of Argentina, Bolivia, Brazil, Paraguay, Uruguay, and Venezuela—further undermined regional exports.[99]

For Mexico, the steady deterioration of exports endangered an essential source of film revenues. In so doing, it weakened the only nearby foreign market large enough to support the big-budget productions that could compete at home with American movies.[100] Without exports, then, the industry was virtually inviable. Just as the U.S. Supreme Court insisted on a postwar dismantling of domestic theater ownership by the American majors—ending the studio system in the United States—the loss of reliable foreign markets meant the end of Mexico's early success in film production and trade. Consequently, and not unlike its Middle Eastern counterpart, the industry spiraled downward in a vicious circle, as declining production and export cut into revenues, undermined investor confidence, and destroyed audience loyalty.

1970s–1990s: Merger and Globalization

The changing nature of competition in the film trade brought weighty consequences to the film world by the mid to late 1960s. Largely as a result of a continuing decline in profits, the major American studios began to merge with, or were taken over by, other business enterprises in this period. In the three decades following Gulf & Western's acquisition of Paramount Studios in 1966, all the other majors either expanded their activities, were acquired by larger entertainment interests, or became part of diversified multinational conglomerates.[101] Competition from other leisure activities presented a greater threat to the U.S. industry than any foreign rival, and the exigencies of risk reduction therefore mandated alignment and recombination with other kinds of economic organization and activity. After hitting rock bottom in the early 1970s, the industry reconstituted itself on more solid ground by forging new horizontal and vertical links to allied businesses.[102]

While the internationalization of the industry had begun years earlier, merger activities contributed to a more complete globalization of the film sector by diluting the national affiliations of firms and strengthening international trade, financial, and cultural flows. Firms of American origin still dominated the industry at the creative level, with Hollywood retaining its status as the film capital of the world and continuing to serve as a magnet for international talent. Yet little else remained fully bound within the confines of individual states. Financing came from increasingly diverse international sources, including foreign state subsidies, European private lenders and, in some cases, the studios' parent companies in Japan, Canada, or

Australia.[103] Markets were substantially integrated for the major American distributors, with foreign earnings equaling or exceeding domestic ones and international coproductions assuring wide U.S. distribution of selected foreign films. Labor moved freely, if often unidirectionally, toward the industry's Los Angeles hub, especially for workers in high-skill positions.

Shadowing the industry's globalization but dating to the early 1950s, the production process itself underwent a critical transformation in the United States, with implications for Hollywood's overseas involvement. American filmmaking, after its original industrial consolidation in the 1920s, had been dominated by fully self-contained and vertically integrated film factories—studios—that engaged in Fordist mass production based on a division of labor. In the postwar era, the industry transitioned to a less hierarchical, more decentralized "package-unit" production system in which sets of firms with specialized knowledge came together for single projects.[104] Independent production companies began to proliferate in the 1960s, many of them filming on location throughout the world and working with other newly specialized companies. By the 1970s, the transformation culminated in the emergence of flexible specialization as the dominant mode of production in the United States.[105] Relatively small independent production companies, formed originally in response to the cost-cutting efforts of the major studios, came to account for a majority of production activity by the middle of the 1970s.[106] The distributors retained great clout as film finance centers and global marketing networks, while a more fragmented, market-driven and short-term-oriented productive structure reconstituted what had been a declining industry, matching developments in other sectors of the world economy.

The foreign trade implications of the rise of flexibly specialized American film production are of vital importance. A globally integrated industry, with independent companies making a large percentage of films, permitted greater dispersal of production activities in the world, favoring anywhere that local economic conditions were advantageous to producers. In this sense, producers outside the United States were structurally comparable to the American independents in their relationship to the distributors, for they also usually lacked complex organizational apparatuses to sell to global markets. Nothing, however, prevented such firms from entering into mutually beneficial business relationships with the major distributors. Many attempted to do so, since the distributors profited simply by marketing successful films, regardless of where they were made. In this way, the seemingly contradictory tendencies of a typical global industry came into being: relatively small, geographically dispersed, and locally based producers maintained ties to world markets through their connections to powerful, integrated, and internationally oriented distribution and marketing firms.

In an integrating global film market, the evolving system of flexible specialization increased the competition facing all firms worldwide. As Michael Storper notes, flexible specialization was not necessarily the only or the optimal mode of production for filmmaking; it was the product of path dependent postwar choices that ultimately defeated the alternatives in the United States.[107] Yet, its success meant that state-run or hierarchically organized firms eventually were competitively disadvantaged and less successful, not for being less effective in filmmaking, but because they were incompatible with the growing, globally integrated distribution networks organized largely by Hollywood. They were mismatched with the latter as institutions of cultural production that had not, and perhaps could not have, evolved in ways similar to what came into existence in the United States. In the absence of a better "institutional fit," and without being able to perform for themselves all of the functions of distribution—effective national, regional, and global sales of popular products—these producers faced a decline that in many cases prompted demands for protection. Whether they got that protection was another matter, dependent on a set of domestic determinants to be considered shortly.

In a broad sense, the globalization of the industry occurred in both of its dimensions: the economic one, involving the film trade as a business like many others; and the cultural one, reflecting the symbolic significance of film as a public medium. Globalizing developments in the two spheres were closely related. While filmmaking was a business activity for its dominant commercial players, and economic criteria prevailed in their decision making, the political context shaping economic choices affected profoundly the nature of cultural production. The influence held by economically powerful American-based producers over issues of content and meaning had real cultural consequences. These consequences were least easily discernable when viewed from a perspective that took American world cultural dominance for granted. On the periphery, however, it looked very different.

The Egyptian experience is telling in this regard. In the first half of the 1970s, Hollywood's presence in Egypt continued a slow two-decade contraction, with American exports declining even more than those of other countries. The American share in the Egyptian market dropped to its lowest point in the mid-1970s, when U.S. exports were less than a third of what they had been late in the 1950s and accounted for only about half the foreign films on the market.[108] This pattern mirrored larger trends in the financial health of both the U.S. industry and its Egyptian counterpart, where the total size of the market for motion pictures had been waning since its postwar peak. The film trade also paralleled broader developments in Egyptian–American relations. Bilateral ties between the two countries had deteriorated in the 1950s after a brief, Suez-induced improvement, remaining formally correct but tense for the rest of the Nasser regime. Anwar Sadat's overture to Washington changed both regional and international equations in the aftermath of the 1973 war with Israel, and U.S. film exports rose then in a manner slightly suggestive of the freeing of political constraints.

The 1980s were marked by a particularly erratic course of expanding and contracting foreign imports, but by the 1990s the American industry seemed to have won back some of the relative dominance it enjoyed in earlier years, partly through the decline of European production. Still, U.S. involvement was at a much lower overall level than its apogee four decades prior. With demand undermined by a shrinking exhibition infrastructure and a stagnant economy, Egypt had declined to the status of a minor foreign market for U.S. exports, with no real role as an American entrée to the regional trade for Egyptian facilities or personnel. Nearly five decades after the famed American director, Orson Welles, contracted with Studio al-Ahram to shoot two different films in that studio and on location in Egypt, American producers rarely sought to brave Egypt's bureaucratic entanglements to produce there.[109]

Egyptian exports to the rest of the Middle East showed three distinctive trends in this late period. First, throughout much of the 1970s, exports rose sharply, particularly in the traditionally strong regional markets of Lebanon, Syria, Jordan, and Iraq.[110] A reduction of Egyptian state control over the industry in 1972 combined with a partial easing of general economic constraints to create new production and export opportunities. Gauging the real strength of the 1970s private-sector export resurgence is difficult, but every indication is that trade volumes were greater than any seen since the nationalization of much of the industry a decade earlier. In all likelihood, recorded trade values have been underestimated for this period, given the strong disincentives for entrepreneurs to report all their activities to the tax authorities and the growth of video piracy as a new problem plaguing the Egyptian industry.[111]

A second but countervailing trend started after President Anwar Sadat's November 1977 trip to Jerusalem and the subsequent Egyptian-Israeli peace process and accord, when the rejectionist states of Algeria, Iraq, Libya, South Yemen, and Syria began boycotting Egyptian film exports.[112] Soon thereafter, moreover, exports to Lebanon were decimated by its civil war, representing a further politically driven diminution of an historically strong export area. Fortunately for the industry, exports to Saudi Arabia and the smaller Gulf states began to rise at nearly the same time, forestalling a crisis and shifting the marketing focus of distributors and producers. Despite the longstanding absence of public movie houses in Saudi Arabia, the Saudi elite had always imported films for private consumption, as well as for foreign workers in the oil industry and to air on Saudi television after 1966.[113] By the 1980s, the relative importance of the Saudi market grew immensely, especially for films made directly for television or formatted for video. Saudi investment in production financing also expanded noticeably in this period, influencing the choice of thematic elements in Egyptian films so much that it limited the appeal of exports in other markets.

Finally, a decline in Egyptian film exports occurred after 1986, by which time the Gulf states had come to account for the overwhelming majority of sales. The drop in world oil prices in the 1980s led to a previously unknown fiscal austerity in the region that curtailed import expenditures. This pattern was reinforced by the Gulf crisis and war of 1990–91, which led to even tighter spending by the Gulf states and eliminated Egyptian exports to Iraq, even if the Syrian and Lebanese markets were reopened. The growth of satellite and other new media delivery technologies further reduced demand for exports, especially to the Gulf states. For their part, North African markets remained dominated by European, American, and Indian competition, though Egyptian videocassette exports made some progress there.[114]

If the Egyptian market suffered from instability and transformation, Mexico's experience was equally turbulent. American exports to Mexico remained dominant in the early 1970s, reportedly accounting for 5 percent of all U.S. foreign sales in 1971.[115] The U.S. hold on the Mexican market had been sliding steadily for years in comparison to Latin American and European exporters. But the United States still exerted great competitive pressure on the domestic industry by virtue of the box office draw of its best pictures, its distribution power within Mexico, and the expectation of continually improving technical quality that its glossy productions created. Even as Hollywood passed through one of the weakest periods in its history, it bested local producers with audiences that had long since grown weary of what had become, according to film scholar Charles Ramírez Berg, Mexico's creatively exhausted and decrepit cinema.[116]

Beginning in 1970 with the *sexenio* of Luis Echeverría, the Mexican state virtually nationalized the industry, causing a rapid decline in private production, from which U.S. and other exporters benefited by expanding sales. This expansion reversed the longstanding numerical decline in American exports. Only in the early 1980s, when the more commercially oriented private sector was encouraged again by Echeverría's successor, José López Portillo, did the U.S. export boom come to an end. At the same time, however, Mexico became increasingly popular among American producers as an inexpensive foreign production site, hosting films like David Lynch's *Dune* and John Huston's *Under the Volcano*.[117] American film exports followed a relatively steady course throughout the 1990s, though the production crisis accompanying Mexico's latest peso devaluation undermined Mexican buying power for imports. Wealthier segments of Mexican society turned increasingly to new technologies to access American films via nontraditional exhibition windows like the satellite dish and cable television.

While president Echeverría had a number of likely objectives in intervening so extensively in the film industry, one of them was to regain Mexico's international markets by returning to the production of socially engaged, better-quality films.[118] Reflecting his aspiration to international leadership in the developing world, he sought to recapture some of Mexico's former cultural prestige and repair relations with the country's intellectual elite, damaged immeasurably by the Tlatelolco student massacre of October 1968.[119] Despite these efforts, the call for a quality-led cinematic renewal met with only limited success internationally. By the mid-1970s, Mexican film exports remained confined to several small countries in Central America and the Caribbean, as well as Colombia, Ecuador, Uruguay, and Venezuela. In most of Latin America, the United States held three quarters of the market, and Mexico competed with Argentina, Spain and a handful of other European producers for a small share of the remainder.[120]

With the return of rampant commercialism under López Portillo, foreign receipts expanded reportedly by more than a third, as production rose to 114 films in 1979, its highest point in two decades.[121] Unlike past growth in foreign sales, these were mostly low-budget potboilers in familiar genres that sold fairly well to popular classes throughout Latin America. Aided in the 1980s, moreover, by the growing export strength of television programming produced by the Televisa conglomerate, Mexican films also grew in popularity in the expanding Spanish-language exhibition circuits of the American Southwest. The United States, with its twenty-five million Hispanics, represented a lucrative potential market that had been underexploited since the early 1950s. The industry's dynamics, however, highlighted the dilemma for state elites faced with a choice between economically successful but culturally vacuous film production on the one hand, and a commercially inviable quality cinema on the other. Most industry leaders and state officials chose the former.[122]

The Limits of Competition

A review of nearly a century of competition in the world film trade reveals four patterns that are significant to understanding how states respond to the competitive pressures on their industries. A first pattern involves the increasing globalization of the industry, a process that has concretized since the 1970s, with consequences for national policy responses. The film trade's early internationalization and the profound importance of foreign markets to Hollywood are difficult to exaggerate. But the worldwide integration of the industry over the past few decades has called into question the very concepts of "national" and "foreign" film production and trade. Non-American markets do not simply represent additional profit to Hollywood firms; they have become integral to the entire structure of production. Without them, not only would large-budget pictures not be made, but the industry as it is presently constituted would fail. State responses to global competition no longer can be targeted easily toward a foreign rival, since domestic interests are so fully invested in the culture trade.

A second pattern is evident in the relative constancy of American dominance in the film trade of the past century, seldom failing to exert powerful competitive pressures on film industries worldwide. U.S. dominance has waxed and waned in particular markets, both at home and abroad. In this sense, the film industry is a distinctively American enterprise of the twentieth century, and U.S. power in motion pictures is emblematic of American dominance in the world: while the industry had the material capacity to do so slightly sooner, it arrived only after World War I; it reached its box office pinnacle in 1946; it saw its lowest point domestically in the last days of Vietnam; by the 1980s it was undergoing a revival; and in the 1990s it began to reinvent itself to accommodate new markets and technologies. Today it is witnessing changes that will likely transform it into an important segment of a globally oriented culture and information industry in the twenty-first century. These developments notwithstanding, the relatively unchanging nature of U.S. dominance in film suggests that the international power of the American industry is insufficient to account for the marked oscillation in state policies worldwide.

A third pattern concerns the importance of the national-level factors that have served as an impetus to international change. Hollywood's frequent depiction as a monolithic and omnipotent world cultural menace belies the actual financial vulnerability of many of its firms, especially in the postwar era after court-mandated enforcement of antitrust laws raised the level of domestic competition among the majors. Many of these firms have remained unaltered in name, but each one has undergone repeated organizational and personnel changes induced by the financial pressure to stay profitable in an extraordinarily competitive and risky business. These changes have affected film industries everywhere, as U.S. firms have been inspired by their domestic competitors to expand their pursuit of international markets and—especially overseas—have been empowered by the U.S. political process to do so in the collusive fashion of a legally sanctioned cartel.

A final pattern relates to the importance of film exports to most national industries. For film production to compete in the home market with the most powerful of the world's producers, large-scale investments are required in both production budgets and related human and physical capital, ranging from film schools to studios and technical facilities. Low-budget, locally made filmmaking is very difficult to undertake as a successful commercial activity by small producers; the availability to audiences of expensively made alternatives, with admission costing the same, undercuts such efforts. Extensive investments in a filmmaking infrastructure can be sustained only by reaching larger audiences than those found in domestic markets alone. Midsized producers with small home markets, such as Egypt and Mexico, have found themselves in a perpetual state of crisis since the shrinking of their exports after World War II. With even their domestic exhibition circuits contracting over time, despite rapid population growth, a downward spiral in film exports has been difficult to reverse. State intervention to bear some of the costs has proved inadequate in some cases and politically unsustainable in others.

The forces of international competition detailed in this chapter play an influential role in the formulation of trade and cultural policy, affecting local film industries in several possible ways. Increasing competition impinges on the market share enjoyed by domestic firms, influencing the profitability and popularity of their films. Competition cuts into the availability of exhibition outlets for locally made pictures, preventing them from reaching audiences and making it more difficult for producers to recoup investments. In regional markets, competition from large, international producers reduces the export earnings of smaller, locally made films, trimming earnings that are vital to amortizing production costs in small markets. Competition affects tax revenues, depending on the structure of the tax law, by increasing revenues when the tax yield of foreign films is heavier than that of local films, or decreasing them if local films attract larger audiences. Finally, international competition also influences the substantive film tastes of audiences, whose subsequent expectations for domestic films are altered by their exposure to the stylistic conventions and high production values of foreign pictures.

The latter could have several possible effects on state policy. Fierce competition for market share and exhibition outlets could lead domestic producers and distributors to press for protective trade barriers when it threatens to reduce their earnings substantially. By the same token, declining foreign competition could undermine the willingness of domestic firms to engage in lobbying, particularly when the costs of political action exceed any possible benefits from a policy change. Heavy competition also could provoke a concerted state policy response if it undercuts the tax revenues paid by local producers, though the opposite reaction could be elicited if high tariffs on foreign films yield greater revenue in conjunction with growing import activity. Finally, under certain circumstances, policy makers could react negatively to the presence of extensive foreign cultural material in the local market.

While the level of international competition changes the policy-making environment in vital ways, these changes are only part of the policy-making equation. It may not matter, for example, that competition causes lobbying by domestic firms. What matters are the circumstances under which such lobbying is likely to yield results. Such circumstances are determined by domestic factors like market structure, which only sometimes give the industry the power to act in a concerted and unified way. It may only matter, moreover, that competition erodes tax revenues when such considerations are an important part of the state's financial calculus, and when tax authorities have some say over relevant policy. Finally, competition may bring foreign cultural influence, but this affects policy only when appropriate state authorities see such material as antithetical to well-articulated cultural goals. In all of these instances, certain logically prior questions must be answered to explain the impact of foreign competition, and these questions cannot be answered without reference to domestic political and economic configurations.

In this sense, a distinction must be drawn between the influence of foreign competition on the film market and its consequences for state policy. What competition can do is not the same as what it can explain. Competition can drive other firms out of business, but it cannot explain the responses of state policy makers to such events. It changes the environment in which policy choices are made, but such choices remain underdetermined. Competition alone does not dictate the structure of the domestic market or the shape of state institutions, since only domestic political actors can formulate the laws determining such structures and shapes, and these actors are subject to competing, local influences. While globalization pressures determine the external circumstances under which choices are made, they play out in a domestic context that accounts best for the choices themselves.

4

Markets and Trade Policy

Egypt and Mexico were subject to similar competitive pressures over time, but their trade and cultural responses differed substantially. Understanding the structure of film markets helps to explain these differences better than the fact of competition itself. In both Egypt and Mexico, the nature of relations among producers, distributors, and exhibitors affected each group's ability to secure desired policy outcomes. In this sense, trade policy reflected the organizational strength and unity of the subsectors, which shaped their capacity to deploy resources in support of their policy preferences. After a brief review of definitions and conceptual matters, the chapter offers a detailed discussion of Egyptian and Mexican film markets and trade policy.

Market Structure

The structure of the market subject to international competition explains the nature of the response to such pressures. Market structure in general determines how the relationship among firms (a structure) affects the generation of prices (a market). Conventional definitions include three elements: 1) the number of firms in a given sector; 2) the ease with which new firms can enter into the field; and 3) the number of consumers they serve.[1] In their varied forms, market structures range from monopolistic to competitive. On one end of the spectrum, a pure monopoly has only one firm, is very difficult for new companies to enter, and serves a large number of consumers. Its opposite, a perfectly competitive market, is composed hypothetically of many firms, can be entered into easily, and serves relatively fewer consumers. With fewer consumers, firms in competitive markets are less able to ignore consumer preferences and must set prices with an eye on their rivals. Firms in monopolies, in contrast, deal with a larger number of consumers, who face greater collective action problems and are

Figure 4.1 Political Consequences of Industrial Organization

MARKET STRUCTURE	POLITICAL CAPACITY
Monopoly Few firms Difficult to enter Many consumers	Strong
Competition Many firms Easy to enter Few consumers	Weak

less able to act in concert. Other things being equal, these companies can set higher prices.

Various market structures have highly significant political implications. Firms in monopolies wield greater political power by virtue of the linkages among them, which facilitate their efforts to lobby policy makers. Markets that are more competitive yield a less politically influential sector, in general, since differences in subsectoral interests hinder cooperation on policy matters. This holds true even in nondemocracies, where firms seek to protect themselves by engaging actively in whatever political process exists to articulate and defend their interests. Even centrally planned economies share a similar dynamic, since—obvious differences aside—state-owned enterprises behave somewhat like monopolies under capitalism. In general, political outcomes result from the interaction of state authorities and social actors—firms embedded in market structures, in this case—all of which seek to secure the most favorable outcomes possible for themselves (see Figure 4.1).

Measuring market structure in the film sector poses relatively few difficulties and can be achieved by analyzing the industry's producers, distributors, and exhibitors. The number of firms in a given market structure corresponds to the number of film producers and distributors. When a small number of producers and distributors have a large share of annual film production, this embodies a monopolistic tendency, typified by the American industry from the early 1920s to at least 1948. Growing monopoly is also readily apparent when producers and distributors join together to form film combines, reducing the number of individual companies. While the ease of entry for new firms is more difficult to quantify, it can be gauged by the amount of capital investment required by a given sector, with more capital-intensive industries typically proving more difficult to enter. This attribute changes infrequently and can be incorporated into an

Table 4.1 Market Structure: Values and Indicators in the Film Sector

Value	Number of Firms	Ease of Entry	Number of Consumers
Monopolistic	Few Producers Few Distributors	High Capital Needs Many Legal Barriers	Many Consumers (Cinemas)
Competitive	Many Producers Many Distributors	Low Capital Needs Few Legal Barriers	Few Consumers (Cinemas)

understanding of the market structure by noting changes in investment requirements for filmmaking.

Although the number of consumers in a given market is gauged most precisely by ticket sales, a simple and reliable proxy for this aspect of market structure is the number of existing cinemas: a market with many consumers requires many cinemas; one with few consumers requires relatively few. Over time, an increase in the number of consumers tends to stimulate the building of new cinemas, just as a decline in the number of consumers eventually leads to a reduction in the number of cinemas. Accordingly, a declining number of movie houses reflects rising market competitiveness, as producers and distributors compete for fewer and fewer outlets (and buyers) for their products. Conversely and counterintuitively, a rising number of cinemas strengthens the monopolistic aspect of the larger market structure, as producers and distributors no longer have to worry as much about having exhibition outlets and consumers.

In short, market structure combines the number of producers and distributors with the ease of entry for the sector and the inverse of the number of cinemas. This is summarized in Table 4.1.

Trade Policy

Like more conventional goods and services, trade policy in film falls along a continuous spectrum ranging from liberal to protectionist. A liberal policy entails minimal government intervention, few import or export restrictions, relatively low tariffs, and the imposition of only minor regulatory burdens on the industry. The state in such cases acts merely to minimize barriers to exchange and protects property rights by combating piracy. Individual consumer preferences are the most significant determinants of import and export volumes, with little formal account given to the national identity of film producers or distributors. A protectionist policy, in contrast, is any government policy that raises the price of film imports or lowers the price of exports. This may occur by the direct imposition of tariffs, licensing requirements, or foreign exchange controls, or by more indirect means, such as the weak enforcement of intellectual property laws. In such cases, trade flows reflect state preferences directly in proportion to the extent of government intervention.

Trade policy in film is evident in relevant legislation, the major policy declarations of state decision makers, and the implementing actions of the bureaucracy, both nominal and de facto. Ten specific indicators are detailed in Chapter 2: 1) tariff levels, 2) import quotas, 3) remittance restrictions, 4) import licenses, 5) foreign exchange controls, 6) intellectual property protection, 7) direct foreign investment, 8) joint ownership, 9) box-office taxes on imports, and 10) subsidies for domestic producers. A broad approach to discerning policy is essential because the film trade, as with other commodities, often displays contradictory tendencies, such as when the state bureaucracy relaxes import restrictions but increases tariffs in response to a foreign-exchange shortage. For this reason, relatively clear policy patterns emerge only over an extended period. In the remainder of the chapter, I review market structure and its effects on trade policymaking in Egypt and then in Mexico, clarifying these dynamics and demonstrating the logic of the argument.

Egypt

The Egyptian film market contained a broad array of countervailing tendencies from the 1920s onward, reflecting the contending interests of its various subsectors, as well as the influence of their foreign partners and rivals. These subsectors—production, distribution, and exhibition—grew at different rates, were subject to different financial exigencies, and had inherently conflicting interests by virtue of their positions in the industry. At times, the fragmentation of interests within and among the subsectors prevented cooperation in the absence of a monopolistic market structure. Cinema owners, for example, could not easily join producers to press for protection because many theaters showed at least some non-Egyptian films. Producers and distributors, only some of whom also coproduced or distributed foreign movies, had a similar difficulty cooperating. Egypt's film sector was replete with industrial conflict, efforts to overcome it, and the effects of interest fragmentation and unification on trade policy outcomes.

Because of these dynamics, Egypt's trade policy in film varied enormously from the industry's very beginning and not always in ways that reflect rising global pressures. Two broad shifts are evident. While trade policy was liberal early in the century, it became quite protectionist in the mid- to late 1950s, just before the general transformation of the Egyptian

Table 4.2 Market Structure and Trade Policy in Egypt

	1930s-1940s	1950s-1960s	1970s-1990s
Market Structure	More Competitive	More Monopolistic	Mixed
Trade Policy	More Liberal	More Protectionist	Mixed

economy wrought by the Socialist Decrees of the early 1960s. It changed again in the 1970s, when trade policy became mixed and limited liberal measures were reinstated alongside protectionist ones (see Table 4.2). Policy changes did not mirror international trends. While global pressures increased steadily, trade policy responses did not move in lockstep with them, sometimes even contradicting the general tendencies displayed internationally. Nor did trade policy simply reflect changes in political regime, since similar policies persisted across the transitions from one to the next.

The industry's early market structure was competitive in 1925, when Egypt's first major production company was founded. The Misr Company for the Theater and Cinema led a small but devoted group of local producers and distributors, who came to match the steady growth in cinemas.[2] Earlier in the century, Italian and other resident Europeans were responsible for most local filmmaking, producing a modest total output of about 150 short films and newsreels.[3] A number of cinemas were established to serve resident foreign nationals. The first of them showed French and Italian movies in Cairo and Alexandria just before World War I.[4] When the war cut off foreign imports, an Alexandrian photographer and a group of Italian businessmen formed a local company to make short films, though local production remained minimal for a decade.[5] After the mid-1920s, however, and the creation of the Misr Company, producers and distributors launched the first efforts by Egyptian filmmakers to shape the context in which they operated, as they began to lobby state authorities to maintain policies that would further their economic interests. From the very outset, the local industry was unmistakably commercial in orientation, developing a genre and star system that mirrored Hollywood and churned out melodramatic and musical productions.

Consequently, the market became even more competitive throughout the 1930s and most of the 1940s, as new companies were established in a rapidly expanding industry. This trend was accompanied by the rapid growth of moviegoing as a pastime for lower and middle class Egyptians, who were drawn in large numbers to the increasing number of Arabic-language cinemas after the 1931 production of Mary Queeny's *Song of the Heart* (*Unshudat al-Fu'ad*). The fastest expansion in the period came during

World War II, when declining foreign imports and rising wages for workers led to an unprecedented growth in production and in moviegoing by Egyptians, only to contract somewhat in the war's aftermath.[6] Most of the producers in this period made only one or two films per year, revealing the disorganized, almost speculative, nature of early production and its financing.[7] Individual producers themselves undertook most of their own domestic and foreign distribution. A few of them, such as Togo Mizrahi and Studio Misr, led the way by their consistent efforts to make and distribute several films each year. But much smaller operators attempted the vast majority of production, either entering the business as a temporary venture or failing to last more than a short while.[8]

Early trade policy in film had a clear liberal orientation that emanated from the European-dominated free trade regime in place in Egypt since the nineteenth century. Until 1930, negligible restrictions were placed on film imports, which were brought in primarily for the large communities of resident foreign nationals. Even after Egypt obtained tariff autonomy in 1930 with the expiration of the last of its long-standing agreements with the European powers, film tariff levels remained relatively low. A few exceptions to the liberal tendency did appear in the 1930s and 1940s. World War II, for example, led to an increase in customs duties on film, which doubled to £E 5 per kilo from 1941 to 1945, as well as an augmentation of the ad valorem duty from 3 to 7 percent. Still, these were relatively minor developments in the face of a decidedly liberal trend, marked above all by very few restrictions on film imports. This liberal course was maintained until the mid-1940s.[9]

Unfortunately, for the industry, the level of competition translated into political weakness for its members. At the height of the postwar boom in 1947, nearly two thirds of Egypt's cinemas still showed either all or some foreign films, assuring that most exhibitors were unwilling to support any degree of control over the level of imports.[10] Without a network of ownership or organizational connections linking producers and distributors to each other and to their exhibition venues, and with barely enough capital to operate on their own, the film sector's weakness nearly spelled its early demise just after World War II. In August 1947, however, the Ministry of Commerce and Industry sanctioned and funded the formation of a Chamber of the Cinema Industry (*Ghurfa sina'at al-sinima*) under the supervision of the Federation of Egyptian Industries.[11] Established in accordance with the law governing industrial chambers, the Cinema Chamber had branches for each subsector, as well as for the film studios and laboratories.[12] It was designed explicitly to solve the coordination problems among producers, distributors, and theater owners, and the Ministry eventually made membership mandatory.[13] While it provided only a weak link among subsectors, its oversight activity and corporatist structure did raise the cost of entry into the sector for new firms, and its establishment marked the beginning of a shift in the industry toward greater monopoly.[14]

As the market connections began to grow after World War II, trade policy took a slight protectionist turn, with the imposition of remittance restrictions, import licensing requirements, and other targeted quotas and duties. The industry came out of the war in an extraordinarily strong position financially, though structural problems in the Egyptian economy limited the gains that filmmakers could realize. On the one hand, production costs in the immediate postwar period averaged £E 25,000 for films that would then earn nearly £E 100,000 at home and abroad.[15] On the other hand, the hard currency reserves accrued during the war began declining rapidly, limiting the import activities of local distributors. Since film imports to Egypt—as in Britain and much of Europe—were not deemed sufficiently important by authorities to warrant the hard currency expenditure, officials put remittance restrictions on imports. In time, certain additional measures indicated a further movement toward protectionism, including the imposition of import licensing, as the state extended its authority into new areas like film. Furthermore, the Egyptian government in the early 1950s placed a three-film maximum on the number of pictures that could be dubbed into Arabic each year, and it raised customs duties on foreign films by 25 percent.[16]

Significant remnants of past liberalism nonetheless remained in the early 1950s. While groups like the Cinema Chamber sought relief from foreign competition repeatedly, they were not usually satisfied. The Cinema Chamber submitted a report to the Ministry of Commerce and Industry in early 1955 that called for several strongly protectionist measures to combat the local success of American blockbusters like *Quo Vadis*. An influential critic of the report, however, declared that "competition is the heart of commerce" and noted the importance of foreign production to the local exhibition market, as well as to tax and customs revenues. Indeed, tax rates were relatively high, as state authorities looked to the entertainment industry for revenue more than anything else, and this was reflected in the level and variety of taxes imposed on all segments of the sector. Still, this did little to restrict the flow of foreign pictures into the country.[17]

The number of cinemas grew rapidly in 1950s Egypt, reaching an all-time high of nearly four hundred late in the decade, though weak seasonal demand occasionally forced some large cinemas to close for short periods.[18] In contrast, the number of producers and distributors, which had exploded just after the war, grew only erratically in the subsequent decade.[19]

Production itself remained fairly disorganized and ad hoc.[20] Union activity by filmmakers intensified and became more influential after the 1952 revolution, making it slightly more difficult for new companies to form and therefore strengthening those that already existed.[21] Rising production costs, moreover, increased capitalization requirements for new companies. This pattern of growing exhibition, stronger unions, mounting costs, and uneven production and distribution translated into an increasingly monopolistic market, allowing the most established firms to operate as they pleased. They did just that, churning out a mixed bag of winning but often formulaic pictures that solidified the rigid star system.[22] Over time, it also gave greater political influence to pro-protectionist forces, with the Cinema Chamber demanding and obtaining a tax increase on foreign films in 1955, claiming that producers and distributors had inadequate access to theaters for the first run of their pictures.[23]

The growth of monopoly, however, was not without serious subsectoral conflict. Theater owners suffered under the weight of a growing tax burden and were the most vulnerable segment in the production and marketing cycle due to the fixed nature of their investments.[24] The state levied five different taxes, with an effective tax rate ranging from 30 to 72 percent, depending on the price of the ticket.[25] Exhibitors attempted to strike back by showing multiple films during peak moviegoing periods, lowering admissions prices, and in some instances, forming cinema chains to control the degree of competition.[26] This undercut the revenues of individual producers, who slowed their activities considerably in the mid-1950s, virtually halting production for four months in a move that pitted them against the unemployed workers of the Technicians Union in 1956. Distributors needing quick revenue sold new Egyptian films to second-run exhibitors shortly after releasing them at first-run movie houses, leading the public to avoid the more expensive theaters and prompting criticism from both exhibitors and producers.[27] The Cinema Chamber itself proposed to establish a specific distribution quota, which would have given each of the largest fifteen distributors the right to sell a certain number of films.[28]

All parties appealed to the state for relief, demanding stricter regulation of the industry through a more comprehensive cinema law, with producers seeking greater controls on foreign imports. Beginning in 1956, the state began to respond to demands for assistance, making the first of several efforts to bolster the ailing industry by establishing a Cinema Support Fund, raising customs duties by an additional 25 percent, and charging foreign films a release tax.[29] A ministerial ruling the following year established a release tax of £E 150 per film for the censorship of all feature-length imports.[30] Rising monopoly in the film sector yielded greater pressure for protectionism, as the number of active producers and distributors declined, cinemas increased, and entry into the field grew more difficult. This culminated in the nationalization of some of the leading producers, distributors, exhibition circuits, studios, and laboratories in the beginning of the 1960s.[31] Clearly, much larger forces were at play in Nasser's Egypt, and film monopoly did not, in itself, purchase this outcome by virtue of the power of filmmakers to bind themselves together and insist on change. But initial state support for the sector did precede the larger structural changes made in the Egyptian economy in the early 1960s, and the evolving monopolistic market presented greater state intervention and a protectionist trade policy as its best option.

The nationalization of key segments of the industry early the next decade restricted imports of motion pictures enough to bring about a contraction of the film market. Implementation of this new protectionism was far from perfect, but the evasion of import restrictions by U.S. firms would not have yielded significant returns in a declining market.[32] As it stood, profits for most imported films were not particularly large, except for an occasional American picture. Even the American majors found themselves cutting back on exports to Egypt, since the restrictive measures assured that marginally successful films simply could not make a profit. By the 1960s, new restrictions on the ownership and operation of foreign companies combined with the cost-cutting efforts of many U.S. firms, leading most of Egypt's foreign distributors to close their Cairo offices and, in many instances, move to more free-wheeling Beirut.

Subsequent developments in the 1960s were a logical consequence of the postwar movement toward greater monopoly. For much of its history, the industry had found itself at odds with the state in many ways, seeing state assistance in general as too little, too late. The creation of the public sector in the early 1960s changed this relationship and reinforced the industry's monopoly power. Nationalization entailed the sequestration of numerous private companies and caused a further decline in the number of active firms in operation, as well as a greater reluctance on the part of new firms to enter the field. A divided system of production and distribution emerged, with a large and dominant public sector responsible for approximately half the film output, as well as for most film distribution and the provision of financing for privately produced films. The public sector also monopolized the pool of resources available to the industry, ranging from studio time to film stock. While private filmmakers never ceased operating throughout the period, many directors and producers found themselves shut out of the system, disconnected from the sources of

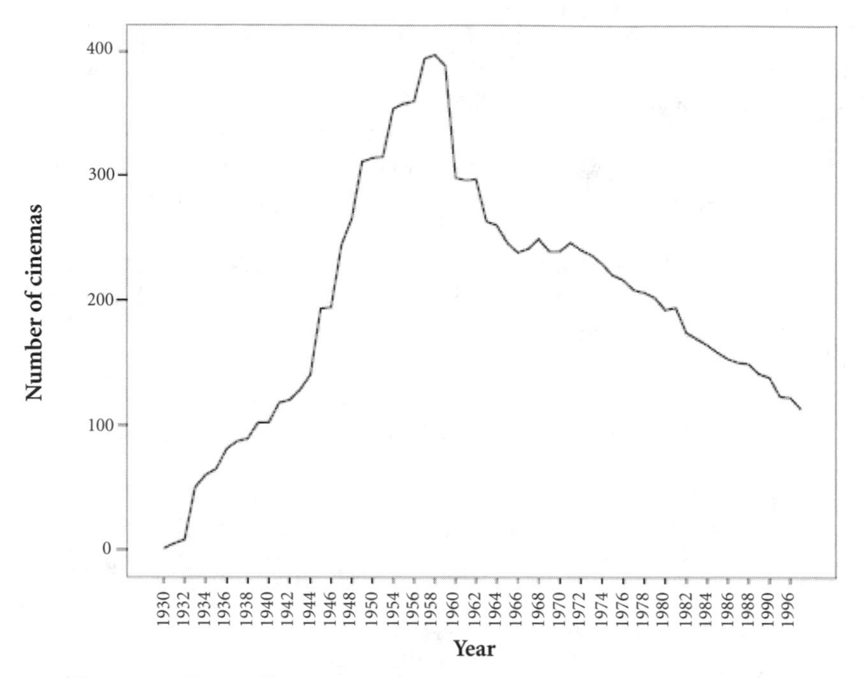

Figure 4.2 Egypt: Cinemas
Source: Compiled from CAPMAS; Ministry of Culture; Cine Film, various years

film financing and studio availability that became part of the public-sector empire of patronage and control.[33]

Running contrary to the rise of the state monopoly, the most important trend in exhibition throughout the 1960s was the beginning of a four-decade decline in the number of cinemas in Egypt (see Figure 4.2).

The state itself took control of about 20 percent of the country's movie theaters, a figure that would rise very slowly to 30 percent in the ensuing decades.[34] Already overtaxed and barely profitable, the liquidation of private cinemas accelerated after July 1960, when the Ministry of Information began providing two channels of television service. Particularly in rural areas outside Cairo and Alexandria, television cut severely into the cinema's potential audience, as it had in the United States a decade and a half earlier.[35] Average annual moviegoing rates for Egyptians, already on the decline, dropped from three and a half times per year in 1960, to two and a half times per year in 1966, to once every three years subsequently.[36]

Egyptians went from patronizing privately made movies to watching public-sector television, transferring to the state the costs of public entertainment and compounding the expense of the state's simultaneous entry into the filmmaking business. Just as the state's tax policies and its introduction of television had sabotaged the exhibition market, the existence of fewer and fewer cinemas made it more difficult for public-sector production to succeed. At the same time, an important programming change in movie theaters shored up the political unity of the state monopoly: while in the 1950s, an overwhelming majority of cinemas ran at least some foreign films, by the mid-1960s, only 40 percent of theaters had foreign programming.[37] This decline of the mixed-program cinema weakened one of the most significant voices in the industry with an interest in low tariffs and a liberal film trade. The only other such voice came from the politically marginal foreign distributors selling American and European pictures in Egypt, and the latter already were dwindling in number. Programming changes in this period therefore meant that foreign distributors no longer had a substantial set of domestic allies in the theater owners showing their films.

The structure of film markets began to change more markedly even before the end of Nasser's rule, driven in part by the tumultuous political impact of the 1967 war. In the late 1960s, the number of producers and distributors rose sharply, some of them making films imbued with a greater social realism than what had been promoted by the ostensibly progressive state sector.[38] This occurred when independent filmmakers from a new generation started renting equipment from small, specialized companies in order to film on location, rather than attempting to work in the poorly equipped but expensive state-owned studios of the public sector.[39] Ironically, the rigidity of the state's monopoly on resources combined with the critical, doubt-filled atmosphere to stimulate the rebirth of competition late in the Nasser era.

The shift to a more competitive market structure continued after Anwar Sadat's consolidation of power in the early 1970s. Amid accusations of fiscal mismanagement and heavy losses in the public sector, the state stopped sponsoring new film production in 1973. Private sector production accelerated throughout the decade, driven by increased foreign demand in the oil-rich Persian Gulf states and financial-sector liberalization that made investment capital more readily available after 1974. Domestic consumption declined even further with the steady reduction in cinemas that became particularly pronounced after 1975, when rising real estate values induced a sell-off of theaters. High ticket prices further undermined domestic demand, as did a growing popular sense in the 1970s that moviegoing was no longer a suitable family activity. The latter sentiment accompanied Egypt's post-1967 religious revival, with Islamist activists voicing renewed criticism of the cinema's contribution to a perceived social decay.

A major shift in film trade policy began in the 1970s, as the Egyptian state withdrew from direct involvement in film production by the public sector and introduced a liberal dimension—albeit modest—to trade policy. With Egypt's economic orientation changing under the Sadat regime and its socialist experiment in decline, the state retreated substantially from active engagement in all fields of cultural production. State policy shifted toward encouraging the private sector, offering the industry only minimally supportive incentives to invest. These changes aside, the film industry in the Sadat era remained mired in a morass of state regulation and oversight. Perhaps to make up for the state's retreat from its earlier activism, protectionist policies grew more severe in some areas. While remittance restrictions remained in place, import licenses were put in the hands of Egyptian producers, who were allowed by law to bring three films into the country for every Arabic-language picture they made. The government also established in 1973 a numerical quota of three hundred films per year that could be imported.[40] The Sadat regime throughout the 1970s thereby withdrew from the role of film producer, distributor, and financier but did not entirely eliminate past protectionism. Trade policy became mixed, despite a clear shift in broader economic policies.

The structure of film markets grew increasingly competitive into the 1980s, powered largely by booming exports of low-budget, made-for-video film and television production destined for the Gulf states. The foreign market affected local production in a substantive sense by boosting socially conservative and entertainment-oriented film content, and economically, by providing a source of sales revenue, employment, and financing in the aftermath of the retreat of the public sector. The majority of film companies—seventy-five out of the eighty-nine that existed in 1985, for example—were small, fly-by-night operations that produced only one picture each year or folded after a very short time.[41] These firms thus were unable to cooperate and demand a protectionist trade policy that would have enabled them to attain a greater share of the domestic market.

State policy under Hosni Mubarak in the 1980s remained relatively unchanged at first, though a growing indifference to the economic health of the domestic industry became apparent. The import quota of three hundred films per year remained in place, but it had little real effect on the activities of foreign suppliers, who had long since recognized the limited capacity of the Egyptian market to absorb foreign films and shifted their attention elsewhere. Tax discrimination targeting foreign companies also continued in the form of the entertainment tax, which was two to three times higher for foreign films than for their domestic counterparts.[42] In fact, state tax policy treated the industry more like a source of revenue than anything else. According to one observer, tax collectors were sent to cinemas every single day of the week to collect the state's share of receipts, including on weekends.[43] All told, customs duties and other taxes were levied reportedly at a rate approximating 80 percent of the value of any given film import.[44]

By the late 1980s, declining demand precipitated another cycle of crisis and took its toll on the number of producers and distributors. At the heart of the crisis was the ongoing shortage of domestic cinemas, which, historically, had supplied more than half the industry's revenues. Some pictures were delayed from domestic release by as much as two years, for lack of exhibition outlets. Within a decade, there were fewer cinemas in Egypt than there had been before World War II, sixty years earlier.[45] Just as American planners in the 1950s and 1960s created suburban multiplexes to replace high-profile, but failing, downtown cinemas, Egyptian entrepreneurs built new and expensive multiplexes in greater Cairo in the 1990s, designed to supplant the large and distinguished theaters concentrated for decades in the downtown area.[46] Not surprisingly, however, foreign imports dominated these cinemas, which catered to Egypt's wealthy and Western-oriented political and economic elite.[47] Such developments only benefited the individual investors serving this small segment of the population.

The most liberal aspects of trade policy in this period were those subject to the strictures of international financial agreements. Such liberalization measures included a relaxation of remittance restrictions, after a 1990 regulation by the Ministry of Economy, and the substantial strengthening of intellectual property laws.[48] Having joined the Berne Convention in March 1989, Egypt had to adopt a stronger copyright law in June 1992, which increased penalties for violations. Piracy in the film sector had first become a substantial problem in the 1970s, when unlicensed Egyptian film and television programming began appearing on television and videocassettes distributed in the Gulf states. With the rise of video technology, American producers and distributors also voiced widespread concern over piracy. But in recent years, losses by U.S. companies in the film sector were estimated to have declined from forty-two million dollars annually to eleven million dollars.[49] Toward this end, a private anti-piracy organization, the Motion Picture Association in Egypt, was established in May 1995, led by General Hussein Hassan Abd el-Rahman, former head of the Ministry of Interior's vice squad.[50]

Market structure in Egypt has seen a fluctuation between competition and monopoly, as various subsectors have alternated between conflict and cooperation with each other in pursuit of their ends. When interests have been fragmented, the subsectors have sought contradictory policies, and this has perpetuated industrial division. Fragmentation prevented, for example, those cinemas that showed foreign films in the 1950s from joining domestic producers and distributors to press for change. By contrast, when interests have been unified, the subsectors have been joined by ties of ownership or organizational connection, and this has strengthened industrial unity. The growth of the public sector in the 1960s, for example, allowed filmmakers, technicians, and workers from all branches to unite behind state-owned enterprises, even if other developments eventually undermined the viability of the industry.

In accordance with these changes, film trade policy has shifted from relatively liberal beginnings to strong protectionism to a more complex and internally inconsistent policy mix. Policy makers sometimes have encouraged the free flow of cultural production, and at other times have restricted severely the entry of foreign films. At any given moment in recent years, moreover, some policies have been protectionist and others more liberal. This ebb and flow, as well as the contradictions of policy choices, do not match the gradual increase in global pressures described in Chapter 3. Nor do they reflect changes in domestic political regimes, another potentially powerful factor often used to explain policy responses to globalization.[51] It is true that trade policy became protectionist in the postwar period and during Nasser's nationalist regime in the 1950s and 1960s. But what is perhaps surprising is that it did not become even more protectionist at an earlier stage, given the regime's political orientation. Throughout most of the 1950s, after all, the state tolerated extensive foreign competition even in the face of vocal complaints by domestic actors. Furthermore, Egypt's film trade policy since the 1970s has been even more puzzling, as it has embodied both liberal and protectionist tendencies. This policy mix has persisted across changes in state leadership, and it has not been as evident in most other sectors of the Egyptian economy. These mixed results are perhaps the most telling, for they reflect the underdetermining impact of international competition on a complex and multifaceted industry that encompasses contending interests.

Mexico

Throughout its history, all the subsectors of the Mexican film industry were politically active, though this activism did not always win the desired results. The relative political strength of producers, distributors, and exhibitors waxed and waned, just as their interests and objectives differed and they were affected in distinctive ways by broader trends in the industry. Differences in the structure of the film market had crucial consequences in this regard. Without formal ties of ownership, and therefore common interests, cooperation among the subsectors was very difficult. The existence at various times of a large number of U.S.-owned cinemas in Mexico, for example, prevented exhibitors from cooperating with those producers who favored high, protective tariffs. In contrast, forceful and unified action by the industry's labor unions sometimes strengthened monopolistic tendencies in the market and contributed to the industry's capacity to achieve desired policy ends. In this manner, conflict and cooperation in the industry was vitally important, as the structure of the market dictated the nature and strength of the political demands made on state policy makers.

Mexico's trade policy in film varied as much as Egypt's, though its policy trajectory over several decades was quite different. After a liberal beginning, policy toward the film trade was heavily protectionist throughout much of the 1930s and 1940s. This was followed by a lengthy period of mixed policies that included distinct elements of both liberalism and protectionism, reflecting the incapacity of the industry to unite in support of a long-term and mutually beneficial trade and industrial strategy. After a last gasp of protectionism in the early 1970s that culminated years of state support, Mexico's film trade policy began to be liberalized. This policy oscillation was too uneven to be a result of steadily mounting global pressures throughout the century, nor did the film trade simply echo the periodic changes of administration or deeper, permanent changes in the objectives pursued by the ruling party. As Table 4.3 shows, trade policy reflected changes in market structure, embodying all of the industry's successes and failures to overcome the inherently conflictual interests driving each individual subsector.

The Mexican film market developed "backwards" in a sense, with the birth and growth of exhibition preceding the emergence of strong local production and distribution.[52] In the first decade of the twentieth century, itinerant filmmakers wandered the country, eventually settling down after dozens of theaters were established in Mexico City during the decade-long Revolution that began in 1910. With the founding of Mexico's first studios, *México Cine*, silent film production became a small industry that yielded ninety films by 1929. The industry's small annual output competed with a

Table 4.3 Market Structure and Trade Policy in Mexico

	1930s-1940s	1950s-1960s	1970s-1990s
Market Structure	More Monopolistic	Mixed	More Competitive
Trade Policy	More Protectionist	Mixed	More Liberal

much larger volume of American imports, though the introduction of sound late in the decade created the possibility of expansion, both domestically and abroad. In the early 1930s, Mexico's half-dozen producers began to increase at a rapid rate and reached twenty-one by 1934.[53] The market thus remained somewhat competitive, since any new Mexican producers joined a large number of strong and relatively unregulated foreign firms to vie for a relatively small domestic audience of sixteen million people, two-thirds of whom were poor and illiterate.[54]

Initially, trade policy in film was fairly liberal, despite the disputes of the 1920s over the influx of American pictures and their stereotyped characterization of Mexican nationals. Admittedly, a discriminatory entertainment tax was imposed as early as 1920, whereby domestic entertainers were charged at a rate of 2 percent and foreign ones at 10 percent.[55] But American pictures soon constituted the overwhelming majority of films released each year and saw few constraints in Mexico's growing market. Just as broader political efforts were oriented toward restoring social equilibrium and creating the political institutions of the modern postrevolutionary state, early economic policy focused mainly on developing the leading sectors of the economy in the 1920s and 1930s. Mexican state officials noticed the film sector's growth and initial success, but this prompted little active involvement from a leadership more concerned with other issues and under no pressure to intervene.

The real surge in monopoly began around 1936, a year that started with uncertainty for the many new producers who had entered the business with the advent of sound filmmaking. Some industry observers feared the impending demise of all national production, since Mexican pictures had not yet proved successful with Spanish-speaking audiences at home or abroad. But the 1936 production of *Over on the Big Ranch* (*Allá en el Rancho Grande*) set the stage for the establishment of a large-scale film industry after its unprecedented international success at the box office and in the Venice Film Festival. This success convinced investors of the industry's great commercial potential and spurred a rush of investment, as production companies sought to duplicate the success of *Rancho Grande* by copying it without even the slightest innovation in the film's content. The ensuing growth of Mexican production in the late 1930s was aided by the virtual disappearance from the market of Spanish-language films from the United States, as well as by the effects of Spain's civil war on its film exports to Mexico. Mexico's chief regional rival, Argentina, produced technically superior products that were strongly influenced by a European aesthetic. Yet Mexico's films were more resonant with local "folkloric" traditions, often had strong nationalist themes, and therefore tended to appeal—at least for a time—to what President Lázaro Cárdenas called the popular classes.[56]

Monopoly was evident because the number of new films greatly exceeded, in relative terms, the number of new producers. This gave existing companies a stronger voice in intra-industry discussions. For its part, the rapidly growing local audience assured strong receipts at the box office, as more people flocked to see a new and uniquely Mexican genre: the *comedia ranchera*. This had the effect of allowing producers to continue making, and remaking, essentially the same pictures, with little immediate fear of audience dissatisfaction. The uncertain availability of exhibition outlets made producers nervous and especially risk averse, but by placing the popular new *ranchera* genre at center stage, producers were able to lessen that uncertainty, maintain a hold on the market, and compensate for the early absence of a well-developed star system that normally would have reduced their risks.[57] Film-related trade unions, moreover, were consolidated throughout the 1930s, and their resulting struggles with producers drove up filmmaking costs and made it harder for new companies to enter the field.[58] As a result of these changes in the number of firms, consumers, and the ease of entry for new competitors, the sector grew increasingly rigid and monopolistic in this early period.

With the industry's consolidation and the growth of monopoly, a more coherent and protectionist film-trade policy developed under Cárdenas's nationalist regime. Even if the cinema needed little by way of active government support, a protectionist impulse was first manifest in the state's contribution to the making of two pictures, *Nets* (*Redes*) and *Let's Go with Pancho Villa!* (*¡Vámonos con Pancho Villa!*). State interest in the cinema led to discussions on the floor of the Senate in 1937 relating to the proposed creation of a national bank devoted solely to providing credits for the film industry. The initiative failed at first, since state attention was focused on the impending expropriation of the oil sector, but its consideration revealed the perceived importance of the industry in this period.

The 1940s began with attempts to unify the industry even further, amid concerns about sinking quality and growing subsectoral conflict. In March 1940, a mixed commission was formed, composed of producers, distributors, and union representatives. The commission sought to deal with attempts by distributors to send the very worst Mexican films to the very best theaters, thereby assuring the rapid box-office failure of these films and their quick replacement by more lucrative foreign pictures. This practice was most harmful to producers, since the exhibitors profited from showing popular imports. With only three big box office hits that year, however, the reforms that were achieved did not solve producers' problems, particularly in securing adequate financing. Distrust between producers and the UTECM workers' union prevented the two groups from cooperating in this area. When the union offered to provide producers with financing credits, they balked out of fear that this would cede too much control to the union, which was responsible for much of the rising expense of filmmaking in this period.[59]

Indeed, shortly thereafter, under President Manuel ávila Camacho in April 1942, the state founded the Cinema Bank (*Banco Cinematográfico*), the only bank in the world devoted exclusively to the film industry. This bank replaced the Bank of Mexico's dominant film-financing company, the *Financiera de Películas*, providing the most promising production companies with credit lines of up to two million pesos, amortizable over ten years.[60] The Cinema Bank proved highly effective during World War II, a period of moderate protectionism. This period also comprised an economic and cultural highpoint for Mexican filmmaking, coinciding as it did with the reduction of American and European exports to Mexico and Latin America. Aside from the Bank's subsidies, a major form of protectionism in this period consisted of the state's weak enforcement of intellectual property rights. Thus, the cinema industry in 1943 was able to take advantage of confusion over royalty payments and adapt to the screen no fewer than twenty well-known literary works from all over the world.[61] The extraordinary level of U.S. wartime support for the Mexican industry mitigated any need for greater protection.

Market monopoly remained in place throughout World War II, during which the Mexican industry benefited from American aid to the film sector. Perhaps just as importantly, a tacit agreement emerged among film directors, who sought to put forth their best efforts. Supporting them was a powerful group of allied producers that appeared in 1941, led by Jesús Grovas and including companies like Filmex, Films Mundiales, and POSA Films, the latter having made some enormously successful pictures under an exclusive contract with its rising comedy star, Cantinflas.[62] Slightly more concentration of production occurred in 1942, when the number of producers declined at the same time that the volume of total production increased. This growing monopoly was soon furthered by the new state Cinema Bank, which provided one million pesos to establish the Grovas company, called at the time "the most powerful film company in Latin America," in order to finance Mexico's six most famous producers.[63] Grovas, Filmex, Films Mundiales, and CLASA were producing one-fourth of all Mexican films within two years, and the latter two companies merged in 1945.[64]

Not all segments of the industry benefited equally from changes in the postwar economic environment, and subsectoral conflict over the film trade intensified in the postwar years, even if the market remained monopolistic. Conflicts revolved around industry fears of an inevitable U.S. resurgence in Mexico and Latin America. Partly as a result, the worst labor crisis in the industry's history erupted in 1945, when a second major union broke away from the long-established Cinema Industry Workers Union (*Sindicato de Trabajadores del Industría Cinematográfica*—STIC). The new Cinema Production Workers Union (*Sindicato de Trabajadores de la Producción Cinematográfica*—STPC) began with 3000 actors, directors, and artistic personnel working in production and led by Cantinflas. The division between the STIC and STPC unions became an important feature of the industrial landscape, separating production workers from those in distribution and exhibition. The significance of this division was first evident in August 1945, when the STIC launched a strike against Hollywood distributors, asking for a large increase in salaries for its exhibitors. Since STPC-dominated Mexican producers were affected, they joined the U.S. companies and refused to provide films to STIC-affiliated theaters throughout the country. Theaters therefore had to make use of old material and appeal to independent distributors for product to fill their time slots, losing more public support than they gained in higher wages by the time of the strike's resolution in September.[65]

While filmmaking had benefited from new infusions of creative talent each year, with the end of the war and the formation of the STPC, union membership became much more difficult to obtain for newcomers: only one new director was allowed to join in 1945, compared to fourteen in the previous year. For this reason, while divisions between the two labor unions contributed to building walls between the subsectors, the "closed door" nature of the new STPC reinforced the monopoly power of those firms already present. It also had a long-term stultifying effect on the quality of the industry's creative output, and this had tremendous consequences for Mexico's position in the broader film market in the years that followed. Increasingly, the domestic market was becoming segmented along class lines: the largely illiterate popular classes remained heavy consumers of Mexican movies, while the better-educated middle class defected to American imports.[66]

The most powerful exhibition monopoly in Mexico promoted this division. William O. Jenkins, a highly controversial resident American citizen and former U.S. Consul, owned hundreds of cinemas and influenced the development of the postwar market by his reported connections to the state's Cinema Bank.[67] He contributed to the cost-cutting wave in the late 1940s by helping to finance a large number of mediocre productions, known as *churros*, which many insiders saw as the industry's salvation at a time when Mexican filmmaking no longer was favored by the exceptional circumstances of war. Producers made *churros* only for those who could not read subtitles, concluding that Mexico could not compete with foreign imports for the allegiance of the educated public.[68] The unions supported this tendency because it assured at least modest continued employment and reinforced their exclusive position in the industry. Such market segmentation allowed for the sidestepping of international competition by directing production away from mainstream cinema and toward a narrow and loyal, if captive, audience.

The monopolistic trend persisted late into the decade. For example, in 1947 the Cinema Bank founded National Films (*Películas Nacionales*), a state-owned distribution group that brought together some of the largest distributors in the country.[69] Exhibition monopolies also grew stronger, and the number of existing theaters rose to meet increased domestic production and new imports. Within two years, the volume of production rose from double to triple digits for the first time ever, remaining there for over a decade. At the same time, investment in production was slashed mercilessly. Average production costs declined from 648,000 pesos per film in 1945 to 579,000 in 1946, and then 450,000 in 1947.[70] A 90 percent devaluation of the peso in 1948, at a time of inflationary costs and national economic difficulty, meant that the actual decrease in expenditures was much more substantial.[71] By 1949, most movies were made in three weeks, on a budget of only 300,000–400,000 pesos, with no provision for paying royalties to the owners of literary rights and the exclusion of expensive foreign talent from productions that were increasingly oriented to local audiences. The star system remained in place to reduce risk and enhance predictability, but production values were cut in every other way.[72]

The postwar production boom appeared to resemble those of 1938 and 1943, but an atmosphere of crisis pervaded the industry, since most productions were of much lower quality than the films of just a few years earlier, and most foreign markets were rapidly being lost.[73] Demands arose for state intervention, and the president of the National Film Commission (*Comisión Nacional de Cinematografía*) called for legal measures to protect the industry and expand both national and foreign markets. Still, the high volume of production under these conditions had advantages for experienced producers and exhibitors, who were able to use the crisis to accelerate the process of monopolization. In fact, the firms most troubled by the situation were the independent producers, who sought to avoid relying heavily on the Cinema Bank for financing or on *Películas Nacionales* for distribution. Efforts to consolidate the industry were confined initially to the growth of monopolies in each of the subsectors, so that when the head of the producer-dominated STPC union was discovered to have ties to the most powerful exhibitors in 1948, he was removed from his position. But when this happened again in 1949, the union let it be known that the growing power of the monopolies was not damaging to workers' interests.[74]

While trade policy became increasingly protectionist, as the industry lapsed into the first of many crises, protectionist measures proved unable to stem the rising tide of imports or strengthen the industry's crumbling economic base. By 1947, when it became clear that the industry's "Golden Age" was over, the state and business leaders proposed what would become a long series of ineffective initiatives to "save the cinema." The first of these was the Cinema Law of December 1949, which—aside from its promotive cultural dimension—was designed to protect national producers from international competition.[75] Article II, Section 10, of the implementing legislation, for example, established the principle of reciprocity as the basis on which authorities would allow foreign film imports into the country. Unfortunately for producers, however, state authorities in August 1952 granted an *amparo* (or writ of *habeas corpus*) to an alliance of domestic exhibitors and their Hollywood suppliers, effectively undermining this particular aspect of the law and reducing the degree of protection shielding filmmakers from their northern neighbors.[76]

Likewise, a newly created Cinematic Support Commission in the early 1950s concluded that producers needed further legislation to protect their interests. Policy discussions pitted the representatives of producers and exhibitors against each other, and an intense debate in the Mexican Senate in October 1952 ended with the supporters of state intervention triumphing over the defenders of classical economic liberalism.[77] The result was a set of amendments to the previous Cinema Law that began with a declaration of the public interest in cinema, and identified the mediation of subsectoral disputes as an important new function for the National Council of Cinematic Art (*Consejo Nacional de Arte Cinematográfico*). The new rules were designed to offer greater protection to producers, even including provisions to enhance state support for the building of new production facilities.[78]

In the early 1950s, the broader structure of the market finally began to change, with some parts of the sector remaining monopolistic and others becoming more competitive. On the one hand, the number of firms declined further, and it continued to be difficult for new firms to enter the business. On the other hand, the shrinking number of consumers—caused by deteriorating film quality—created stronger competition. By 1952, producers found themselves increasingly disadvantaged vis-à-vis the Jenkins exhibition monopoly, and they sought state protection. Exhibitors like Jenkins derived much more revenue from foreign pictures than domestic ones, therefore opposing any protection of the local industry and growing even stronger as one independent theater after another sold out to him. Consequently, by 1954, more movies were being made than exhibited, and for several subsequent years only about 80 percent of new films actually made it to the screen. In some instances, this led producers to invest their Cinema Bank credits in other, more profitable activities in order to repay them.[79] This practice was only curtailed in 1955, when the Cinema Bank stopped lending to producers directly and began funneling loans through the three leading distributors.[80]

Indeed, the state reorganized the financial and distribution structures of the film sector in 1954, bringing together the three large, publicly regulated distribution organizations: *Películas Nacionales* (covering national distribution), *Películas Mexicanas* (for distribution in Latin America, Spain, and Portugal), and *Cinematográfica Mexicana Exportadora*—Cimex (the rest of the world). The Cinema Bank henceforth provided up to $700,000 per film to these distributors, which passed the financing along to the individual producers of their choice. Significantly, however, the distributors themselves were owned partly by groups of production companies, with a state ownership share of 10 to 40 percent, according to the head of the Cinema Bank.[81] Both profits and losses in all three distribution companies were borne by their constituent member firms in proportion to the gross earnings of each movie they supported, rather than their overall capital holdings.[82] Consequently, the distributors only funded producers they considered highly reliable, and no producer who wished to keep operating could afford to take chances. In effect, this created a rigid network of commercial patronage that protected powerful companies and assured their continued operation.

Nonetheless, state policy toward the film trade was less than coherent, consisting of a mix of liberal and protective measures for the remainder of the 1950s and 1960s. The new regime of Adolfo Ruiz Cortinez in 1953 called for further measures to return the industry to a sound footing and reinvigorate exports, and his director of the Cinema Bank, Eduardo Garduño, devised an ambitious plan. The Garduño Plan was intended to result in "intense and preferential" protection of Mexican films.[83] It called for assuring that all national films were distributed at home and abroad, with the aim of recovering more fully the investments made by producers and weakening the monopolies enjoyed by foreign-allied exhibitors. The Plan's protectionist implications were substantial, since it limited to 150 the annual number of import permits to be granted for foreign films, with further import constraints to be based on the all-important principle of trade reciprocity. It also provided for large tax-generated production subsidies and called for the creation of a set of distribution organizations, to be composed exclusively of Mexican production companies that were filming in the country and using only Mexican labor authorized by the unions.

For all its ambition, the Plan's actual effects were much less dramatic, as the country's powerful and independent exhibitors found ways to exceed the volume of permissible imports. Still further reforms were undertaken, largely involving the consolidation of state financing companies with leading private distributors into a revamped Cinema Bank and three large distributors. In a 1955 interview, Director Garduño indicated that the state sought only a basic measure of trade reciprocity on the part of countries like Italy and France, which had all but stopped importing Mexican pictures the previous year.[84] At the same time, an increasingly influential middle class in Mexico wanted a more austere cinema of social melodrama from countries precisely like Italy and France, films free of the tacky vice that had become prominent in local production of the postwar years.[85]

In addition to the emphasis on reciprocity, trade policy in the mid-1950s was designed to give producers an advantage that would allow them to compete with the best foreign rivals. The Cinema Bank, for example, funneled millions of pesos into financing expensive "super-productions" that used color cinematography and wide-screen processes like CinemaScope and SuperScope.[86] After the Cinema Bank obtained thirty-five million pesos in financing from the Department of Treasury, the Bank of Mexico, and the state development bank, it reduced loan interest rates for producers from 12 to 10 percent.[87] While the bank had funded the producers themselves in earlier years, by 1955 it directed all of its resources to the three major distribution companies. The latter provided from 60 to 85 percent of a given film's costs, depending on a subjective assessment of its likely commercial success.[88] This arrangement created an environment in which the most established domestic producers were favored.

By the late 1950s, the state was granting extraordinary privileges to the film sector. Through the Cinema Bank, it provided unprecedented financial subsidies that were the envy of Mexican industry. Credits from the Cinema Bank accounted for an exceedingly large share of investment capital, and these credits were all that prevented a complete collapse of domestic production. Producers' profits generally were not reinvested in filmmaking precisely because there was no need to do so, with a seemingly endless stream of state financing available to those who were sufficiently entrenched in the system.[89] Such easy access to state financing prevented otherwise insolvent producers from going bankrupt, while masking the extent of the broader crisis, as Mexican films lost both foreign and domestic markets.

As subsectoral conflict generated more liberal trade policy, producers in the late 1950s complained of a lack of direct state protection and the high cost of doing business. Toward the end of the decade, such costs led a group of producers to reopen the *América* studios in order to employ cheaper

STIC personnel to make thirty-minute short films, ostensibly for television. These films were strung together and sold to theaters as serialized features, undermining the position of rival STPC workers. Such collusion between producers, the STIC, and the exhibition monopolies generated vehement complaints by the STPC, which launched a paralyzing strike in August 1957.[90] But the total volume of production remained high, achieving a record 135 pictures in 1958.[91] A countervailing monopolistic trend precluded firms from falling by the wayside, since new credits were always available for continued work. It also prevented new talent from entering the business. The number of new directors remained very low throughout the postwar era, and the average number of films each of them made rose and fell only with changes in the total volume of production.[92]

By the early 1960s, the industry found itself in an odd position that resulted in an unusual pattern in trade policy. Widespread government activity in support of the industry was not matched by extensive restrictions on foreign pictures, which made their way to Mexican screens and retained the loyalty of middle class audiences. Film policy therefore consisted of a form of toothless protectionism alongside trade liberalism, serving only the interests of a small group of well-positioned industry insiders. The administration of Gustavo Díaz Ordaz inherited this inconsistency in 1964 and attempted to reorganize the industry. The Administrative Council of the Cinema Bank led the initiative. It emphasized in its public pronouncements the economic importance of filmmaking as a source of employment and as one of Mexico's only national industries that exported completely finished products, earned foreign currency abroad, and lessened the depletion of domestic hard-currency reserves by taking up screen time in cinemas all over the country.[93] The new head of the Cinema Bank, Emilio Rabasa, also proclaimed the need for reform and instituted changes in the way the three major distributors granted credits to producers. Díaz Ordaz appointed future president Luis Echeverría to head the Interior Department, and Echeverría was hailed as a "natural leader of the Mexican cinema" for being the younger brother of a film director.[94]

Throughout the early to mid-1960s, another serious crisis—this one perhaps even more threatening than the last—prompted greater subsectoral conflict, even as the volume of film production fell to its lowest level in over a dozen years. The state, for its part, was fully complicit in this downward spiral. State policy makers were not able to withdraw support from an activity many considered important, nor could they find a legislative solution to the sector's continued difficulties, despite repeated attempts. Director Garduño of the Cinema Bank became quite dictatorial at the height of the crisis in 1961, only exacerbating the situation.[95] The result was ongoing but ineffective state intervention, ranging from the purchase of exhibition venues to the consideration of more protectionist legislation and the purchase of studios.

In time it became evident, that most state initiatives were vaguely formulated calls for improved cooperation across the subsectors. In the end, virtually no fundamental changes were realized, and state policy continued to assure the survival of an otherwise ruined industry by supporting its most essential segments while not restricting foreign imports. For the rest of the decade, facing a nationalist upsurge and extreme crisis in other areas of Mexican national life, film policy was of little importance to Díaz Ordaz. Trade policy remained an uneasy mix of protectionism and liberalism, reflecting the cooperation engendered by the state monopoly and the conflict of private interests in exhibition and distribution.

In an effort to confront the problem of theatrical monopolies, for example, the state bought the Jenkins circuit of 365 cinemas in 1960, hoping to assure greater access to exhibition venues for locally made pictures. It did so, however, at a time when only 5 percent of Mexican movies reportedly were successful at the box office. This meant that the various cinema assets it had come to control were endowed with strongly conflicting interests: films that were produced by the state or financed by the Cinema Bank were now losing money in state-owned cinemas that could have operated more profitably by showing only foreign pictures. Before the exhibition takeover, the Jenkins monopoly was blamed for denying access to Mexican pictures, but now there was no such excuse. Audiences certainly could not be forced into theaters. The tensions that this generated ramified throughout the industry, as other elements joined with one segment of the state sector or another to defend their particular interests.[96] The seemingly endless cycle of conflict prevented the various subsectors from acting in unison.

While producers welcomed the closed nature of the unions—a closed shop reduced uncertainty and stabilized earnings—they found themselves frequently at odds with workers over wages. These very same producers also invoked the deplorable conditions facing workers in order to request government subsidies, which then would be used to make more low-quality pictures that did little to advance the prospects for the industry and its personnel.[97] STPC workers, for their part, went on strike in July and August 1961, securing by September a significant wage increase of 10 to 23 percent.[98] These benefits were not extended to STIC workers, causing further resentment by the latter and spurring them to make more low-quality serialized films. They made thirty of them in 1963, almost half of the industry's feeble total output in a year in which Mexican filmmakers began traveling elsewhere in Latin America to take advantage of cheaper labor and fewer union regulations.[99] Such tit-for-tat behavior benefited each group momentarily as corporate entities, but it did very little for Mexican filmmaking as a whole. With narrow segments of the industry pursuing their individual interests, cooperation proved impossible, and the result was damaging to them all.

The structure of the market remained unchanged for most of the Díaz Ordaz administration, as the various subsectors continued to clash. While production costs still were rising due to wage increases and the permanent shift to color filmmaking, ticket prices at theaters remained frozen by law at 1952 levels. From each ticket, the earnings breakdown among subsectors approximated the following: producers earned 20 percent, distributors were paid 7 percent, exhibitors kept 54 percent, and 19 percent went to taxes and royalties.[100] This meant that although producers using STPC labor faced higher costs than those hiring STIC workers, they still only were entitled to recoup these costs from a small percentage of ticket sales. With price controls in place, moreover, these yields were shrinking in real terms.

The mixed market structure of the period saw a final resurgence of monopoly in the first half of the 1970s, with a short-lived but significant state monopoly in film, created by the new PRI President, Luis Echeverría. The latter was motivated by a personal agenda involving his wish to repair relations with Mexican intellectuals and youth, which were damaged terribly by his tenure as Minister of Interior during the 1968 student massacres at Tlatelolco. Support for a revitalized film industry also was compatible with his leftist, Third Worldist political views. His personal motives aside, this new policy was the next logical step in more than twenty years of crisis and creeping state interventionism in film. Since the postwar beginning of the industry's structural crisis, the state had intervened repeatedly, if to little salutary effect. Entrenched interests had thwarted important attempts at renewal, and fuller state intervention was perceived to be necessary for change.

In fact, the shift had begun as early as 1968, when the Directors' Section of the STPC started accepting new members, some of whom were recent graduates of Mexico's first major film school, the University Center for Cinema Studies (*Centro Universitario de Estudios Cinematográfico*—CUEC), founded earlier in the decade.[101] In time, these new directors and a new generation of producers began working in public-sector production, making eighteen of the ninety-five films in 1970 to signal the end of the closed-shop nature of the STPC after twenty-five years.[102] The opening of the union represented an unprecedented transformation, though it was induced by the passage of time more than anything else, as older filmmakers began to retire. The change also accompanied the disappearance of qualitative differences between films made by the two unions and occurred as tension between the STPC and STIC finally appeared to be abating.

The president chose his brother, Rodolfo, to head the Cinema Bank, replacing Emilio Rabasa, who was named Mexico's ambassador to the United States. Rodolfo went to great lengths to unify all segments of the industry, claiming he had the support of both unions from his own past involvement in filmmaking.[103] His family ties to President Echeverría gave him unprecedented power that enabled him to operate with impunity, perhaps giving him greater control of the Cinema Bank than his younger brother had of the Mexican state.[104] Rodolfo Echeverría therefore instituted a series of protectionist measures to reinvigorate an industry that was by all accounts dying. Changes included ending the minimum box-office earnings requirements that kept films from running for very long in cinemas, and that had been set at a higher level for Mexican films than for foreign ones. The new rates had the effect of protecting the national industry from foreign competition by permitting longer runs for Mexican pictures.[105] For his part, Luis Echeverría also attempted to raise production financing by promising state support to potential investors in their other business dealings.[106] While the state never resorted to the full range of protectionist measures, more resources were being devoted to domestic filmmaking by the end of Echeverría's term in 1976 than had been the available for decades.

Just as there never had been so many well-trained directors entering the industry at once, the industry's nationalization was decisive in giving filmmakers new resources and confidence. As in other areas of the economy, however, the cooperative nature of relations within the industry did not stem from good will, so much as from the strength of the state monopoly and its alliance with a small number of private-sector groups. Accordingly, the concentration of industrial and financial resources did not preclude private exhibitors and distributors from reaping the benefits of nationalization. As movie ticket prices were unfrozen, and poorly performing second- and third-run cinemas were renovated and upgraded to first-run venues, the earnings of private exhibitors and distributors more than doubled.[107] One particular theater operator controlling 16 percent of the country's cinemas, for example, is said to have earned 60 percent of all national exhibition revenues by owning the best theaters in the capital.[108] Such private companies colluded with the state by being the dominant shareholders in the state distributor, *Películas Nacionales*, and they had nothing to risk because they were relatively uninvolved in production.

In 1977, after nearly three decades of mixed markets, the post-Echeverría era saw the break-up of the state monopolies and the return of domestic competition. While Echeverría's prior commitments meant that state production exceeded private output for one more year, the López Portillo administration began avoiding any new involvement in the film sector immediately after assuming power. In 1977, the first of three state production companies, *Conacite I*, was liquidated, and the traditionally dominant private producers re-entered the field quickly, albeit with mixed results. On the one hand, the winding down of state production permitted a resurgence of low-quality pictures, primarily for export to the U.S. Spanish-speaking market. At the same time, an average of seventeen films each year were made by independent producers during López Portillo's term, introducing an additional element of competition.[109]

This same year, however, represented the last one in which the state would favor Mexican film production, since the new administration of José López Portillo began to undo the efforts of his predecessor. The new president named his sister, Margarita López Portillo, to direct a newly formed state institution: the Directorate of Radio, Television, and Cinema, which took the first steps toward dismantling the state sector. Justifying the policy change on economic grounds, she attempted to bring well-known foreign filmmakers to Mexico to work, even if this yielded few benefits for the local industry.[110] The state Cinema Bank, after decades of existence as the only such entity in the noncommunist world, was closed in 1979. After Mexico's accession to the GATT in 1986, economic policy in general moved toward greater liberalization. With continued funding from the television industry, film production remained substantial throughout the 1980s, contributing to a resurgence of media exports to the United States and Latin America. This included the growth of what one filmmaker has called Mexico's *maquilladora* film industry, in which American producers used Mexican facilities to assemble their products with cheap local labor, repatriating virtually all of the profits from films made locally but sold worldwide.[111]

This renewed competition was furthered by the reemergence of old divisions between the subsectors. The state-owned Theater Operating Company (*Compañía Operadora de Teatros*—COTSA), sought to undermine state productions in the late 1970s because Mexican public-sector films performed so poorly compared to foreign imports. Many state films simply were never shown, but COTSA also worked to justify the industry's denationalization by debuting some state-made movies without any advertisement, on inappropriate dates, and in poor venues, assuring their box office failure. This occurred at a time when the industry had begun to regain public attention through the success of some of its films.[112] The year 1977 thereby proved to be a watershed for the film sector. Before then, Mexican cinema could be understood as a complex and multifaceted national industry. After that year, it was not so much an industry as a tenuously situated set of business activities undertaken by an odd combination of technicians, laborers, artistically oriented individual filmmakers, and commercially driven entrepreneurs.

The film market under Miguel de la Madrid, Ernesto Zedillo, and Vicente Fox remained competitive, although production dropped to its lowest level since World War II. The dismantling of the Cinema Bank in 1979 combined with the national economic crisis in 1982 to eliminate the most significant sources of production financing. The various subsectors, for their part, were too weak and divided to demand state protection in an increasingly global market. The rise of competition from powerful television conglomerates in the 1980s constituted a final and decisive blow, draining resources, interest, and talent from filmmaking. In response to the crisis, many film workers from the STPC simply enlisted with foreign producers renting out Mexican facilities. The long-term rental of Churubusco-Azteca studios in the 1980s by U.S.-based RKO, for example, led many workers to migrate to foreign production companies, settling for continued employment but low wages.[113] While many firms did compete in the Mexican film market of the 1990s, the majority of them were globally integrated multinationals of foreign origin.

At the same time, the state began privatizing and liquidating the last of its holdings in production, distribution, and exhibition. In 1990, it closed its remaining production companies, *Conacine* and *Conacite Dos*, and in 1992, it sold off the COTSA exhibition circuit and began efforts to privatize Churubusco-Azteca studios. COTSA subsequently was broken into three separate circuits of theaters, and in a related initiative, authorities privatized the last two publicly owned television stations in this same period.[114] Not surprisingly, as the state withdrew from heavy involvement in film financing, commercial production plummeted. The number of films made in Mexico dropped by nearly 50 percent from 1989 to 1990, and then by another 30 percent the next year, to be reduced to its lowest level since the 1930s.[115] A 1992 film law took further steps to reduce state intervention in commercial filmmaking required more decreases over time. For much the rest of the decade, the industry produced fewer and fewer pictures each year, releasing only eleven new films in 1995. While filmmaking in Mexico had experienced a seemingly endless string of crises since the late 1940s, for the first time in several decades, observers questioned the likelihood of continued production by locally based Mexican filmmakers. Some of its highest profile actors and directors—Gael García Bernal, Alejandro González Iñárritu, Alfonso Cuarón—made their way to Hollywood.

Conclusions

Four general patterns emerge from the above. First, analytical attention to the politics of subsectoral interaction is essential to understanding policy responses to global pressure. The film industry cannot be regarded as a monolithic whole, with uniform interests, since developments in any of its distinctive subsectors may influence outcomes decisively. In 1960s Egypt, for example, protectionist policies won out because theater owners lost their audiences to television and to a declining standard of living, both of which undermined moviegoing. Shrinking audiences caused a steady decline in the number of cinemas after 1960, and this strengthened protectionist advocates, since the diminishing power of exhibitors undermined the leading voice of liberalism in the film trade. Theater owners were not necessarily liberal out of ideological predisposition, but their interests required it due to the popularity of the foreign movies on which they often relied. The structure of the market thereby shaped subsectoral battles and the industry's capacity to realize policy change.

A second pattern relates to the clear association between subsectoral conflict and liberal trade policies, since competitive markets tended to yield this kind of outcome. In both Egypt and Mexico, when the various subsectors of the industry were at odds for any reason, more liberal trade policies were likely to ensue. This was especially so if such conflict persisted over time and was accompanied by a decision to abandon national films by subsectors with liberal preferences. The political mechanism linking conflict and liberalism is apparent in 1950s Mexico, for example, when a postwar crisis drove a wedge between producers, distributors, and exhibitors, who were unable to cooperate to prevent a resurgence of foreign imports. This led distributors and exhibitors to abandon Mexican pictures and trade in popular American ones. In the face of external pressures for a liberal trade policy, subsectoral conflict meant that national industries were much less able to offer unified resistance.

A third pattern is the obverse of the second: subsectoral cooperation within the industry led to greater protectionism. Such cooperation took various forms, but the easiest to manage was what occurred in monopolistic markets. Monopolies, whether privately owned or run by the state, produced greater cooperation because economic interests were aligned and relationships of hierarchy precluded actors from expressing or acting upon disagreements. In 1960s Egypt, for example, the newly formed state film monopoly brought all sectors into relative harmony and facilitated their cooperation. New restrictions were created to protect the state industry from foreign competition, and the nationalization of leading distributors and exhibitors deprived foreign competitors of domestic allies to lobby for open access to Egyptian markets. When confronting globalization, only monopolies could support the cooperative efforts needed to resist the liberalization of the industry.

A final pattern relates to the effects of more ambiguous market structures on policy outcomes, a frequent reality that defies simple categorization. Egypt and Mexico each experienced lengthy periods when their markets had strong elements of both monopoly and competition. In Egypt, this occurred primarily after the 1970s; in Mexico, it happened from the 1950s to the mid-1970s. Mixed markets led to inconsistent trade policies that entailed a combination of protectionism and liberalism, and such inconsistency sent confused signals to the industry that increased uncertainty. The resulting responses to global pressures were internally contradictory because they reflected a fragmented national market, in which some subsectors gained and others lost. Just as the winners and losers from changing trade patterns are distributed unevenly, the gains and losses from rising global competition are dispersed throughout the field of players in the sector.

The dynamics of the industry's subsectoral politics can be restated in more general terms. When the subsectors were able to solve what amounted to a kind of Prisoner's Dilemma, overcoming their collective action problems in the process, such cooperation produced policy outcomes that were more favorable to them both as distinct corporate units and as an industrial whole. When faced with import pressures from foreign competition, and in the absence of such cooperation, liberal policies prevailed. Liberalism under these circumstances meant the domination of the local industry by powerful foreign rivals, who worked hand in hand with local subsectoral allies. The state sometimes intervened to solve the industry's collective action problems and facilitate cooperation, but often it proved ineffective in this regard. State intervention supporting a public good of national culture worked only when institutional arrangements sought and promoted such intervention.

5

State Institutions and Cultural Policy

Who pays the piper when it comes to national culture: the state (for political purposes) or consumers (for individual benefit)? This question underlies the fundamental tension between state- and market-driven cultural production. Having demonstrated how markets affect the film trade, I now turn to how the state institutional profile influences profoundly the decision to intervene or withdraw in matters of cultural policy. The relative position of the state's cultural and economic institutions usually privileges one set of actors and interests over the other. Following a brief elaboration of the definitions of institutional parity and cultural policy, I review the development of state institutions and detail cultural policy changes in the Egyptian and Mexican cases over the past century.

Institutional Parity

International competition in film is mediated by those state institutions that make and implement economic and cultural policy. The structural relationship *between* such institutions is captured by the concept of institutional parity, which reflects the political equivalence of the state's economic and cultural regulative bodies. The configuration of institutions helps to explain state interests and priorities, with clear policy implications. As Figure 5.1 shows, low parity means that institutions overseeing cultural matters are administratively subordinate and economic interests are dominant, giving less voice to cultural concerns in policy debates. When confronted with even mild international pressure, the state may prove surprisingly acquiescent.

High parity, depicted in Figure 5.2, translates into a stronger, more promotive voice for cultural interests, regardless of the economic cost, since

Figure 5.1 Low Institutional Parity

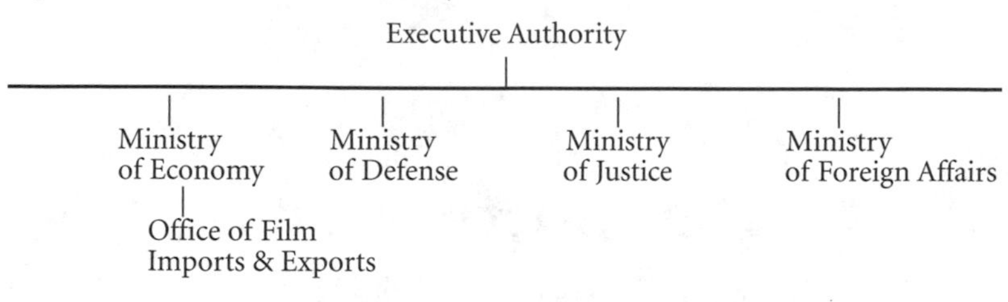

institutions governing the cultural domain are administratively autonomous from their economic counterparts. When facing even severe international competition, the state may show unexpected resistance.[1]

National political leaders are constrained, at least to some extent, by the institutional landscape in which they make, implement, and change policy. Determined leaders may be able to alter the policy-making milieu, but they are seldom free to ignore these constraints, even if political and economic circumstances favor policy reform. This holds true even in authoritarian regimes, where the bureaucratic "tail" often enough wags the political "dog." Ideas about policy options, moreover, are embedded in state institutions, which develop self-perpetuating political supporters with an interest in one kind of policy or another. Officials in state cultural institutions usually advocate an activist role in the oversight of the film sector, while their economic counterparts tend to favor treating filmmaking like all other industries. Much of the institutional story in this account finds high-level leaders wrestling with institutional constraints to achieve their desired policy objectives.

Measuring institutional parity requires analyzing the specific structure of decision-making authority for the film sector. This entails an examination

Figure 5.2 High Institutional Parity

Table 5.1 Institutional Parity in the Film Sector

Value	Dominant Interest	Characteristics
High	Cultural	Organizationally autonomous state cultural institutions
Low	Economic	Organizationally subordinate state cultural institutions

of the formal and informal rules determining which institutions have pride of place in state decision making, and noting significant changes over time.[2] In some cases, state institutions with formal policy-making responsibility delegate authority to other agencies through a bargaining process. In others, bureaucratic conflict subverts formal authority in decision-making turf wars. To determine the primary interests driving policy, a disaggregation of the state reveals all relevant decision-making bodies and their relationship to each other, avoiding tautology by defining the particular institutional configuration independently of its effects.

Cultural Policy

Cultural policy consists of the state's efforts to shape national identity and articulate a public philosophy embodying its most significant values.[3] It targets many areas, some of which have strong commercial dimensions, such as filmmaking, while others are more elusive of direct state regulation, such as the tenor of public discourse on a given matter. State cultural policy seeks to influence, and sometimes to control, the prevailing definition of salient social identities. In so doing, it addresses the self-conception of individuals and their various corporate units by highlighting, defining, identifying, and sometimes obscuring social cleavages. On a practical and daily basis, cultural policy presents and rejects notions of how people should live, work, and pursue their goals, as well as validating and invalidating the goals themselves. It articulates the state's relationship with culture, which can take varied forms.

In the film domain, cultural policy has two possible orientations. The state either promotes its cultural goals directly and actively in filmmaking, or it takes a more indirect, laissez-faire position that delegates authority and influence to dominant societal groups. A promotive state policy deploys a variety of policy instruments to set and realize national cultural goals, such as using screen-time quotas to assure the exhibition of locally made films. Representing a close relationship between the state and culture, a promotive policy also entails strengthening state-sponsored contests, film-related institutions (e.g., technical facilities and training centers), and the state-run censorship regime, which is a system of reward and punishment designed to elicit substantive modifications in cultural content deemed particularly valuable or objectionable by state authorities. More obviously, the state may lead the film production process, promoting certain themes and national images in its cultural output.

A laissez-faire orientation, in contrast, does not constitute an absence of cultural policy, so much as the relinquishing of initiative to nonstate actors. The state does not surrender its capacity to intervene in cultural affairs—no state has ever been entirely absent from this arena—but it allows others to work actively on its behalf. While some level of censorship usually continues to exist, it is more oriented toward excluding specific ideas and images than promoting a particular conception of identity. Such a reluctance to engage in active support of policy goals reflects a satisfaction with the results generated by dominant social and economic forces, whether in the commercial realm of film production, distribution, and exhibition or in the more overtly political domain of censorship and any existing system of state contests and awards. It is a minimalist approach to cultural policy, compared to the maximalism characterizing promotive policies.

Egypt

Twentieth-century Egypt had a rich array of institutions that were active in determining the state's relationship with cultural production. The specific location of governance for filmmaking varied significantly over the years, but it included the Ministry of Culture, the Ministry of Interior, and the Ministry of Trade. These ministries and other government agencies had bureaucratic branches and affiliates devoted exclusively to the film sector. Changes in the general array of institutions were due partly to internal political and economic developments, but institutional changes had persistent and independent effects that defied easy manipulation by state leaders.

The Egyptian cultural policies that emanated from the state institutional landscape varied considerably from the 1930s onward. While policy initially was laissez-faire, it became strongly promotive under Nasser in the late 1950s, and even more so by the 1960s. In the 1970s, state policy shifted back to a largely laissez-faire orientation under Sadat, even if promotive elements reappeared in the Mubarak era (see Table 5.2). These changes can be explained by the degree to which state culture-regulating institutions were accorded bureaucratic equivalence to economic ones. The remainder

Table 5.2 Institutional Parity and Cultural Policy in Egypt

	1930s-1940s	1950s-1960s	1970s-1990s
Institutional Parity	Lower	Higher	Lower
Cultural Policy	More Laissez-faire	More Promotive	More Laissez-faire

of this section traces the development and major shifts in Egyptian cultural policy over the course of the last century.

Institutional parity began very low, early in the century, as state economic actors dominated the decision-making structure. The most important state institutions regulating film were the Ministry of Finance and the Ministry of Interior, the latter's primary concerns relating to neither the economy nor culture. Additional formal authority over cultural policy was divided between the Ministries of Education and Social Affairs, both of which were infused with a strong measure of paternalism that saw the Egyptian masses as needing cultural guidance. A regulation on public establishments in 1911, for example, focused principally on maintaining public order and social standards of propriety, as did Egypt's first censorship law in 1914.[4]

An essentially laissez-faire attitude toward cultural production in pre-Revolutionary Egypt reflected the regime's arms-length approach to culture. No screen quota for Egyptian films existed in the 1930s, although local industrialists proposed one.[5] Throughout the 1930s and 1940s, state authorities believed that while motion pictures played an influential and potentially beneficial role in society, this role was limited mostly to the ethical formation and acculturation of individuals in a general sense. They saw the popular melodramas and musicals of the period as performing a quasi-educational function by adhering to a conservative set of narrative and presentational norms that were agreed upon by consensus and were only occasionally articulated explicitly.[6] Little state intervention was considered necessary to assure compliance with these norms. Beginning in 1933, the Ministry of Education awarded modest cash prizes to individual producers and actors for laudable work.[7] Foreigners, the wealthy, and the landowning elite sponsored high culture, but the state did little to intervene on a national level.

This view of the appropriate state relationship with cultural production also informed the minimal censorship that existed in filmmaking. Admittedly, some close state regulation was evident early on. In its treatment of imports, the Ministry of Finance passed a decree in the early 1920s that restricted imports based on their social content; and in 1928, it began requiring exports to be reviewed by the Interior Ministry, which consolidated its own Office of Censorship of the Press and Publishing in 1936. In 1945, censorship authority was moved to the Ministry of Social Affairs, only to be returned in 1948 with the outbreak of the war in Palestine.[8] In handling exports, moreover, the Ministry of Interior began requiring prior approval almost immediately after Egypt started producing its own films for Arabic speakers in the late 1920s, with the stated purpose of protecting the country's prestige and reputation.[9] But all of these state activities were limited in scope to the restriction of cultural production deemed either unsuitable for popular consumption or deleterious to Egypt's image abroad. Greater involvement was resisted, as one observer of the period has noted, because the Egyptian state was "sufficiently indifferent to culture as to need no policy, administration, or budget."[10]

Throughout the pre-Revolutionary period before 1952, the institutional division of labor was clear in its prioritization of economic matters over cultural ones. With overall direction in the hands of the Ministry of Interior, the Ministry of Finance issued decrees regulating acceptable content for film imports and exports. The Ministry of Social Affairs did seek to improve the quality of cinematic production by sponsoring contests and offering modest prizes to filmmakers, and the Ministry of Education housed a Department of Culture that saw motion pictures as playing a salutary role in the moral and intellectual development of young people. All told, however, filmmaking was considered just another business enterprise, requiring regulation only to the extent to which it brought together the public in ways that potentially could threaten political order and stability.

The Egyptian revolution eventually transformed the state's relationship with culture, though not immediately in 1952, for the Free Officers showed little initial interest in motion pictures. But in 1953 the regime named one of its own, Wing Commander Waguih Abaza, to head a new film production and distribution company, the Nile Company for the Cinema.[11] At the same time, it created the Ministry of National Guidance, headed by controversial writer Fathi Radwan, making an important institutional innovation to facilitate the control of information disseminated by the increasingly prominent instruments of the national media. The new Ministry also established the Arts Agency (*Maslahat al-Funun*) in 1955, led by a prominent writer, Yahya Haqqi, showing sensitivity to the potential influence of the arts in governance.

At first, postrevolutionary Egyptian cultural policy followed closely that of the ancien regime, which had intervened only minimally, even when the industry expanded during and after World War II. The Free Officers set age restrictions on theater attendance by minors in 1954 and revised the censorship law to give a higher priority to maintaining public order in

1955.[12] Filmmaking was viewed by the Nasser regime in populist rather than in high-art terms, with an emphasis on mass entertainment over intellectual expression, social mobilization, or the articulation of a political program.[13] Clear limits prevailed throughout the first half of the 1950s on the amount of assistance available to the film industry, whose proponents resented bitterly the relative lack of attention to their perceived needs and the extent to which bolstering tax receipts seemed to be a top priority in the eyes of authorities.[14]

The rising institutional profile of cultural affairs became more apparent in 1956, when the Department of Culture in the Ministry of Education was transferred to the Ministry of National Guidance. With responsibility for all the arts, literature, and mass media, the Ministry had full authority over the film industry and began to work independently of the other ministries to determine state policy in this area. It sought the advice of a new consultative body, the Supreme Council for the Promotion of the Arts, Literature, and Social Sciences, and both institutions viewed filmmaking in largely noneconomic terms.[15] In essence, this French-style model of the appropriate relationship between the state and culture ruled in Egypt for the next decade and a half.[16]

The watershed in policy development began in 1956. A new cinema law divided the year into three seasons and mandated the exclusive exhibition of Egyptian films during one week each season.[17] Overseen by the Ministry of National Guidance and its Arts Administration, the law represented a first effort to promote national culture on the home front by a regime that would seek increasingly to articulate and deploy identity issues to further its political ends. President Nasser attended regularly the openings of major new Egyptian films and hosted visiting international filmmakers.[18] Limited material incentives were offered in the form of state prizes, and some tax exemptions were granted for films imported as part of cultural and scientific exchange programs.[19]

The film industry gained even stronger institutional patronage in July 1957, when the Ministry of National Guidance formed the Organization for the Support of the Cinema (*Mu'assasa Du'im al-Sinima*). This body took control of most cinema-related state functions and pursued several official objectives, including raising the industry's artistic and professional level, supporting the production and exhibition of politically serious films, awarding state prizes, and promoting the industry at home and abroad.[20] Filmmakers were encouraged by the state's commitment of resources and its consolidation of authority into a single entity that provided them with a dedicated institutional ally. In 1958, as a direct precursor to the formation of the state sector, the Organization for the Support of the Cinema was renamed the General Egyptian Organization for the Cinema, just as the Ministry of National Guidance became the Ministry of Culture and National Guidance. That same year, the Ministry of Finance and Economy agreed not only to grant censorship tax exemptions to films imported for cultural, scientific, or educational purposes, but it transferred future authority in this area to the Ministry of Culture and National Guidance.[21] Long-range and extensive cultural planning began under Tharwat Okasha, the second Minister of Culture, who held the position initially from 1958 to 1962 and sought to support a restricted volume of film production that tapped into Egypt's rich cultural patrimony and Arab identity.[22]

The new Ministry regulated culture and information with a high degree of autonomy in the subsequent decade, and its bureaucrats oversaw a cinematic empire that had tremendous fiscal and creative autonomy. In January 1960, it authorized the General Egyptian Organization for the Cinema to begin participating directly in film production and formed a committee to read scripts and choose the best work for state financing.[23] When the state nationalized Bank Misr later in the year, it acquired Studio Misr, which was subordinated to the Ministry of Culture, along with most of Egypt's other studios. Filmmaking was held partly in limbo for the first two years of the decade, as more cinema activities were appropriated under the auspices of the Socialist Decrees. But few constraints were placed on spending and planning in the next several years. The Cinema Chamber was abolished in 1962, since its commercial orientation as a representative of the private sector fit poorly with the state's new socialist goals. At the same time, Abd al-Qader Hatem took over the Ministry, and in 1963, the state produced its first three films under a consolidated General Egyptian Organization for the Cinema, Broadcasting, and Television.[24]

The political logic of cultural promotion drove state filmmaking from the outset, just as it animated the foray into television broadcasting in 1960. The state companies developed ambitious plans to produce large-scale motion pictures, to publicize Egypt as a major site for international coproduction, and to develop the national infrastructure of studios and theaters. To achieve these ends, the bureaucracy was expanded dramatically. In 1964, the Ministry began building "Cinema City" in Giza, modeled after Rome's *Cinecittà*, and it established an associated, though short-lived, Ministry of Foreign Cultural Relations.[25] Shortly thereafter, the Ministry founded the Technical Center for Arab Cinematic Cooperation to work toward the development of "the Arab film," and it created a research center to determine the educational, scientific, and cultural uses of film.[26] This was followed in 1966 by the formation of a Bureau of Mass Culture to promote cultural goals in the countryside.[27]

State cultural policy became more active, didactic, and highly promotive in the subsequent years of the 1960s, as the regime tried to rearticulate its conception of Egyptian national culture and identity, using the cinema as a partial means of doing so. While he was never so inflexibly committed to socialist cultural ideals as his Soviet counterparts, Nasser's turn to the left in the early 1960s led, nonetheless, to the nationalization of much of the film industry, accentuating the promotional trend in the cultural domain that had begun a few years before.[28] Socialist rhetoric made its way into cultural discourse, finding expression in classic tales of antifeudal struggle like Youssef Chahine's *The Land* (*el-Ard*), along with social comedies like *My Wife, the Director General* (*Mirati, el-Mudira el-'Amma*). Public sector film production grew dramatically in the mid-1960s under Abd al-Qader Hatem, who looked to expand the scope of the ministry's activities while advocating a lighter, more instructive if pedantic cultural policy.[29] The results were highly controversial, with its supporters and detractors disagreeing to this day over the results.

Critical success or failure aside new tensions emerged between the cultural and informational divisions of the Ministry by the mid-1960s. This led President Nasser in 1966 to separate the Ministry of Culture from the Ministry of National Guidance, with the latter retaining responsibility for media activities and the public dissemination of vital information. He brought back Tharwat Okasha as Minister of Culture to contend with the ministry's numerous problems, repair some of the damage, and mollify those who, excluded from participation in the bloated film patronage machine, had left the country to work in Lebanon and France.[30] He also sought to restore the film industry's capacity to compete with rival exports to the Middle East from the United States and India. Okasha found a hugely expanded bureaucracy, serious financial overextension, stagnant studios, and dissent in the ranks of filmmakers and technicians.

Okasha's Ministry proceeded to reorganize the six state companies into two large divisions under a reconstituted General Egyptian Organization for the Cinema, beginning a second phase in public-sector filmmaking that placed new emphasis on cinematic realism.[31] The Cinema Organization at the end of the decade still received one quarter of the entire Ministry budget, second only to archaeological expenditures, and it could claim a number of important critical successes among the films it supported.[32] With the Organization at the helm of its institutional infrastructure, the Ministry founded the Academy of the Arts in 1969 to house all of its arts-related schools, including the state film school, and it set up a National Film Archive in May 1970 to hold Egypt's cultural patrimony in film.[33]

The most important statement of cultural policy late in the Nasserist era came in June 1969, when Minister Okasha spoke before the National Assembly, defending the public's right to participate in cultural activities unhindered by economic constraints and rejecting its commercialization.[34] Okasha maintained that a certain degree of centralized state planning was essential and affirmed what had become a typically paternalistic view of cultural "development." Film policy in Nasserist Egypt thereby embodied what would become a destructive contradiction. On the one hand, the state attempted to use the industry to promote Egyptian and Arab national identity, and to advance a moderate socialist agenda. It did so with the same heavy-handed inefficiency that accompanied the growth of all its burgeoning state-owned enterprises. On the other hand, filmmaking in Egypt had always been organized and understood on a highly commercial basis that flew in the face of direct state control. Moviegoers and filmmakers alike saw the medium as popular entertainment more than education, and this tension between state goals and popular expectations was a major cause of the eventual unraveling of the state cinema.

After Nasser's death in September 1970, the Sadat regime began to make changes in the film-related institutional landscape, initiating eventually a period of lower institutional parity during which economic interests once again displaced state cultural concerns. The new regime immediately changed the name of the Ministry of National Guidance to an only slightly less Orwellian-sounding Ministry of Information, and reorganized the Ministry of Culture by presidential decree the following year.[35] The Ministry of Information remained charged with "ensuring the cultural and political education of listeners and viewers," while the Ministry of Culture had to be satisfied with a shrinking share of a state budget dedicated mainly to economic development and national security expenditures.[36] A year later, in 1971, the state stopped all public-sector film production and merged the Cinema Organization into a new Authority for the Cinema, Theater, and Music.[37]

Cultural policy initially remained unchanged in the early 1970s under Sadat, though state funding for filmmaking began to dry up immediately, with the reorganization of the public sector and the narrowing of priorities to protect Egypt's economic solvency and national security. Still, Sadat appointed an Assistant to the President for Cultural Affairs to manage the cultural component of general development plans, and the Ministry of Culture established a screen-time quota for domestic production. Law No. 13 of 1971 imposed yet another round of mandatory exhibition regulations, again requiring all cinemas to reserve one week per three-month season for the exclusive showing of Egyptian films.[38] Ministerial decrees limited the showing of, and then banned entirely, karate film imports from Hong Kong and Japanese samurai pictures—deemed a pernicious influence on Egyptian youth—as well as Indian films, perhaps considered a box-office threat given their ubiquity and general appeal to audiences.[39] To encourage the efforts of private-sector filmmakers, the Ministry of Culture reinstated the provision of cinema prizes after a hiatus of more than seven years.[40]

State cultural policy in the 1970s embodied all the active, educational, and patronage-providing aims developed in the previous decade.[41] The Ministry maintained an outwardly promotive disposition by dividing the country into three concentric administrative circles: cultural zones, provinces, and centers. The state's paternalism remained unshaken, typified by a June 1972 directive by the Ministry of Culture's censorship department, admonishing Egyptian film distributors to trade only in films that could be seen by people of all ages. As Samir Farid notes, while ratings systems throughout the world came to distinguish between children and adults, Egypt's censorship regime treated all Egyptians like children.[42] Even if state cultural planners no longer had deep pockets, their ambitious aims for remaking the citizenry remained intact, at least nominally.

These changes coincided with an ongoing shift in the relationship between cinema and the state, as the former was distanced from the main propaganda-generating apparatus and left under the control of a revamped but politically marginalized patron of the arts. Accordingly, a promotive cultural policy gradually gave way to one of laissez-faire. As state institutional structures shifted toward supporting entrepreneurial activity, filmmaking became more closely associated with privately produced mass culture than with what remained a fully state-controlled network of radio and television. Since new policy goals included encouraging the private sector, the defunct Cinema Chamber was reconstituted in 1972 under the Federation of Egyptian Industries.[43] Unlike its earlier incarnation in the 1950s, producers in the Chamber were given a dominant voice in its administration, though their demands for assistance fell mostly on deaf ears in subsequent years.[44] Economic interests thus were given pride of place among those state institutions regulating filmmaking.

Accompanying the economic opening, or *infitah*, in the middle of the decade, was a decline and then disappearance of the stridently socialist ideological tone that had energized a promotive cultural policy, which began to return to its laissez-faire roots. The state showed a renewed faith in the power of the private sector. The reinstatement of the Cinema Chamber notwithstanding those who hoped that Egypt's economic opening would provide an investment boost for filmmaking were sorely disappointed. State policy did little to save the crumbling cinema infrastructure of studios and cinemas, even if it declared that the construction of new theaters would be encouraged by law. Developers actually targeted cinemas for destruction despite a 1979 building law intended to promote the inclusion of cinemas in new buildings.

Perhaps in response to the failed promises of the transitional 1970s, social criticism found its way into the cinema. Unanticipated by cultural planners, a new genre of movies appeared, attacking tradespeople and the beneficiaries of the *infitah*—the *munfatihun*—whose newfound wealth made them the object of great popular scorn, especially among Egypt's declining middle-class, low-level bureaucrats with stagnant incomes and new university graduates unable to find professional employment.[45] Reacting to these veiled criticisms of the Sadat regime, Minister of Culture Gamal 'Otayfi issued new censorship rules in April 1976, placing greater restrictions on free expression that were reminiscent of Egypt's strict 1947 regulations.[46] This censorship sought to block political opposition and maintain social control, rather than attempting to create social or national identities, as had earlier uses of the cinema.

In November 1979, moreover, the regime formally abolished the Ministry of Culture, along with supporting agencies that included the Cinema Authority. To compensate for the loss in organizational power, it created in 1980 the Supreme Council for Culture, in addition to a National Film Center and two state companies for cinema-related holdings.[47] The latter two companies were designed to institutionalize a formal separation between the economic and cultural dimensions of filmmaking, with cinemas and distributors sequestered from the ostensibly more creative personnel occupying the studios and production companies. The Supreme Council articulated a new, minimalist cultural policy the following year that included establishing a cinema support fund to raise film quality, assisting worthwhile new projects, and encouraging the industry at home and abroad.[48] In policy planning, it worked in conjunction with the National Council for Culture, the Arts, Literature, and Information, one of the specialized national councils that had begun a series of annual consultative meetings in 1979 and was directly affiliated with the presidency.[49]

Cultural policy shifted subsequently to reflect the state's diminished interest in direct support for filmmaking. In 1980, Egypt's Supreme Council for Culture articulated "the new cultural policy," emphasizing the role that Egyptians themselves should play in cultural development. Consonant with the movement away from pan-Arabism, the regime drew more heavily on Egypt's unique historical lineage than its Arab identity.[50] Most significantly, it called on the cultural and intellectual elite, the *muthaqqafun*, to bear more fully the burden of Egyptian cultural development. It declared that the Ministry of Culture had been designed only to facilitate development in this sphere, not to engage in cultural production. Too many people, it held, looked to the Ministry to act, and stood by without participating themselves. The statement called, moreover, for a closer connection between culture and economic policy, noting that "true culture" in this era was the culture of the people in all its groups and classes.[51]

The film sector of the 1980s under Hosni Mubarak saw few immediate institutional innovations to match or contend with the private sector's wave of cheap, poor-quality video exports to the Gulf, which rose sharply in 1984 and peaked in 1986.[52] The Ministry of Culture, however, was reestablished in the mid-1980s, initially committed to little more than monitoring permits and licenses, as well as to overseeing the Cairo International Film Festival beginning in 1985.[53] The state encouraged the participation of Egyptian intellectuals in cultural regulation, and many remained involved in what became a complex and multi-tiered system of cultural regulation. The state cinema apparatus had seven divisions by the end of the decade: the National Film Center, the Censorship Administration, the agency for "cinema palaces," the Misr production and distribution companies set up in 1980, the state film school, and the Cultural Development Fund created in 1989 to improve cultural services and to plan state participation in funding cultural programs.[54] The many and varied elements of the state cultural bureaucracy, however, had limited resources and narrow regulative authority that focused more on assuring the industry's political loyalty than its cultural direction.

Under Mubarak, the state still tried to shape its citizenry, at least minimally, and to control the discursive cultural arena, but it did so without committing substantial material resources. One of the only efforts to promote Egyptian filmmaking directly was a ministerial decree in 1983 that established new screen-time quotas by requiring the exhibition of Egyptian films during major Muslim holidays, conventionally a time of heavy moviegoing by Egyptian youth.[55] The state also supported a post–Camp David reassertion of Egypt's regional leadership in cultural production, exporting a substantial volume of film and television programming to the Gulf. In so doing, it permitted Egyptian cultural exports to be heavily influenced in substance by these conservative Gulf market destinations and funders.

Likewise, heavy censorship remained in place, even if it was intended to restrict ideas more than shape identity. Political control over cultural production had been easier in the Nasserist era because public-sector filmmakers were either preselected for their favorable political views or they engaged in self-censorship to protect their positions. The freeing of cultural space that accompanied the economic opening of the 1970s led eventually to disputes between filmmakers and state authorities in a time of steadily mounting social tensions, some surfacing in the 1980s. Notable conflicts included those over *The Ogre* (*al-Ghul*), which almost was banned for its purported agitation against the regime, including a final scene that echoed the assassination of late president Sadat; *Door Five* (*Khamsa Bab*), which lost its censorship approval after authorities declared it a threat to public decency and national security; and *The Innocent* (*al-Bari'*), for its depiction of Central Security troops firing on their superiors, anticipating an actual uprising that occurred in 1986.[56]

The degree of parity between cultural and economic institutions changed slightly in the 1990s, after an escalating battle—cultural and otherwise—between the state and a resurgent Islamist opposition. State cultural institutions were strengthened, for example, when decision-making authority for the film trade was returned to the restored Ministry of Culture. Import quotas and decisions regarding the suitability of particular films for export were administered by the Import and Export Committee, formed by the Ministry and overseen by the National Film Center.[57] In general, however, state cultural institutions regulating filmmaking remained subordinate to other institutions and interests. Institutional actors like the Ministry of Information handled the nonmilitary aspects of opposition to the regime, co-opting regulative authority from the Ministry of Culture and perpetuating the latter's remoteness from the regime's political strategies of legitimation and control. The strongest bureaucratic elements in the Egyptian state supported a political role for cultural production, but not one associated with filmmaking.

Indeed, one of the sharpest edges of cultural policy in the 1990s concerned the long-standing conflict between the state and its Islamist opposition, marked by government attempts to define Muslim identity. Cultural conflict between Islamists, the state, and Egyptian intellectuals of various stripes had been simmering for several years before June 1992, when political Islam reemerged from relative dormancy with the assassination of a prominent secular writer. The security services remained the state's primary means of persuasion and social control. But the cultural dimension of this conflict led to state attempts to define and limit Muslim identity by claiming, through definitional fiat, that political dissidents were not truly Muslims and that violence committed by antiregime militants were the acts of non-Muslims. State policy also presented ideas about valid Islamic practice and what it meant to be Muslim, promoting Islamic moderates in the media to both occupy the political center and co-opt an avowedly Islamic vantage point. In advocating this perspective, the regime had the enthusiastic support of some prominent members of the filmmaking community, including leading stars like Adel Imam, whose hit film, *Terrorism and Kebab* (*al-Irhab wal-Kabab*), was awarded a state prize of £E 100,000 for its less-than-subtle condemnation of political Islam.[58] Other filmmakers found themselves drawn into the conflict, targeted by Islamists for their work.[59]

By the end of the 1990s, the regime decided that the best chance of winning its culture wars at home and reasserting its regional dominance abroad was through the deployment of new technology for use in state-controlled film and television production. It built a vast, multimillion-dollar production facility at Giza, called Media Production City, which was designed to enable it to produce large-scale film and television programming for domestic and foreign distribution. To aid the latter, the state also launched its first geosynchronous satellite, Nilesat, facilitating the transmission of Egyptian television throughout the region. Costing millions of dollars at a time of national budgetary difficulty, Nilesat and Media Production City reveal the regime's changing cultural strategy and priorities, indicating that state planners see Egypt as having a comparative advantage in closely managed forms of cultural production, such as television and video. More broadly, Egyptian state cultural policy combined paternalism with a recognition of the role of simple entertainment—bread and circuses—in placating an increasingly alienated youth. This represented a victory for the Ministry of Information's long-standing approach to cultural production. It also entailed a political and economic marginalization of the film industry, except to the extent to which it yielded tax revenues and produced fodder for the information machine.

In responding to competition in the cultural domain, Egyptian state policy has shown some puzzling tendencies. Cultural policy was laissez-faire both early in the century, when global pressures were at their weakest, and in the 1970s, when a mounting challenge from the opposition would have warranted a more forceful cultural response. An understanding of the changing state institutional landscape is useful in illuminating such policies and more so than other domestic or international variables. Even the promotive policy under Nasser did not correspond neatly with the advent of his regime, since promotion was not adopted until Nasser was in power for some years. After Sadat turned to a laissez-faire policy, his regime continued initially in promoting national identity in film. In so doing, the regime pursued policy objectives that made sense only as the persistent legacy of prior institutional arrangements that defined state purposes vis-à-vis cultural production.

Mexico

A limited number of major institutions oversaw the Mexican state's relationship with national cultural production in the last century, including the Department of Interior and several institutions devoted specifically to the film industry. At the turn of the twentieth century, the most active state institutions governing the cultural domain were local in orientation, only gradually becoming subsumed under the national executive authority. During the prerevolutionary regime of Porfirio Díaz, a liberal spirit reigned in cultural matters so that no centralized censorship body existed. City officials in the capital granted exhibition licenses, collected taxes, assured the safety of theaters, and monitored films for moral concerns. State officials increased their oversight in the first years of the revolution, but authority remained heavily localized, focusing more on morality than politics. Filmmakers themselves were strongly influenced by the positivist trend in the intellectual life of the period, directing their efforts toward documenting "the truth" and reality of the revolution. Perhaps inevitably, state authorities were drawn into monitoring the particular version of the truth being recorded, and this led to closer state oversight of the cinema, the institutionalization of national censorship in 1919, and the eventual transformation of the early documentary into narrative fiction filmmaking by 1920.[60]

Mexican cultural policy toward filmmaking varied as much as that of its Middle Eastern counterpart. Just as artistic and cultural life was deeply involved in twentieth-century Mexican national development, state intervention in—and periodic retreat from—this area was a remarkably enduring phenomenon for decades. After an apparent indifference early in the century, state policy in the 1930s and 1940s became strongly promotive of filmmaking, contributing to the development of a pioneering industry in the cinema of the developing world. A more mixed policy emerged in the postwar years, and this trajectory continued until the mid-1970s, when cultural policy took a laissez-faire turn.[61] These changes are summarized in Table 5.3 and detailed below.

As a direct consequence of the privileged early position of the documentary film, the film industry became directly implicated in the nation-building project of modern state elites. Reflecting the progressive ideals embodied in the Constitution of 1917, Mexico's first film-related state institutions were constructed with a concern for the social and political

Table 5.3 Institutional Parity and Cultural Policy in Mexico

	1930s-1940s	1950s-1960s	1970s-1990s
Institutional Parity	Higher	Mixed	Lower
Cultural Policy	More Promotive	Mixed	More Laissez-faire

substance of film more than for its character as an economic enterprise. A high level of institutional parity therefore was evident, so that cultural interests were accorded a privileged place in the state apparatus. State agencies like the Departments of War and Marines, Agriculture and Development, and especially the Department of Public Education saw the new medium as a useful pedagogical device, and they sponsored filmmaking throughout the 1920s.[62] By the time the Cárdenas regime began consolidating the state and building Mexico's modern governing institutions in the latter part of the 1930s, state decision makers had an understanding of filmmaking that was directed toward national educational and cultural goals more than toward promoting the narrative film as a local economic activity or a form of public entertainment. The institutions they created to oversee the industry displayed a high degree of parity, since they were administratively autonomous of state financial concerns.

The first serious effort by the Mexican state to promote the industry was inspired by the invention of sound recording in filmmaking, which gained the attention of state policy makers because of its potential implications for the learning and use of the national language. Before then, the industry had caught the eye of policy makers only as a result of the negative portrayal of Mexicans in American pictures in the early 1920s. Subsequent state sensitivity to a variety of identity issues included a concern over the potential negative effects of film on Mexicans' command of the Spanish language. In 1929, for example, the government of President Emilio Portes Gil issued a decree requiring the "absolute purity" of Spanish translations of foreign film titles, waging a campaign against English-language films and forcing imports to be dubbed into Spanish in 1930. State policy was highly supportive of the development of Spanish-language sound film, even when opposed by advocates of the old silent cinema. In 1930, the government, seeing the mobilizational potential of cinema, obtained a Vitaphone machine from Paramount to record presidential speeches before large crowds at the National Stadium.[63] In 1933, it intervened for the first time to censor a sound film called *The Thirteenth Prisoner*.[64]

More direct forms of cultural promotion began early, compared with state activity by Mexico's Latin American neighbors and the North American model that weighed heavily on the industry. While early state intervention often took the form of financial subsidies that affected trade, some subsidies were linked explicitly to film content, representing the first articulation of state cultural policy. The election of Lázaro Cárdenas in 1934 introduced a left-leaning president who sought to join nationalism with a concern for social issues, and this combination augured well for state use of the film industry as a policy-making instrument. After issuing a decree in January 1935, pledging to lend all possible support to the film industry, the state first assisted a newly founded film company, CLASA (*Cinematográfica Latinoamericana*), which built the most modern studio in Mexico. The Cárdenas regime lent CLASA an extraordinary variety of military equipment for the filming of *Vámonos con Pancho Villa* in 1935, and when the latter picture subsequently almost bankrupted CLASA, the state provided further support by commissioning several short documentaries that were believed to serve the national interest.[65]

Mexican governments were positively inclined toward the sector in these early years, promoting the development of Mexican national culture in film. The Department of Public Education, for its part, helped to finance the making of *Nets* (*Redes*) in 1934, an allegorical tale of struggling Mexican fishermen. It also established a National Film Archive (*Filmoteca Nacional*) in 1936, shortly before the founding of the *Cinémathèque française*, otherwise considered the world's first film archive.[66] Although cultural policy centered on issues of morality and identity in these early years, the question of foreign penetration of the domestic market became significant and led to the first use of screen-time quotas: a presidential decree under Cárdenas in 1939 required all exhibitors to show at least one Mexican film per month, sparking heated debate and an unsuccessful attempt by exhibitors a year later to convince Cárdenas to abrogate the decree.[67] Subsequently, the Camacho government established a national award for the arts, and a National Committee of the Film Industry met in February 1940 to award prizes for the best production.[68] State authorities oversaw another national commission that categorized new films by quality and directed the better movies to better theaters, eliminating altogether the very worst Mexican pictures.[69]

The first and oldest film-related state institution that emerged from 1930s' Mexico was the censorship authority. It was not until April 1941 under President Camacho, when production was expanding rapidly during World War II, that a formal censorship bureau—the Department of Film Supervision—was created in the Department of Interior.[70] While the Department's chief concern was to assure public order, it also played a more unorthodox role in promoting a quality cinema that expressed *mexicanidad*, or a certain image of what it was to be Mexican.[71] As the state's major point of contact with the film sector, the latter institution was overtly political and cultural in focus, and it harnessed the industry to the construction of a national identity that originally had taken form during the Mexican revolution. The censorship bureau focused on perceived external threats to Mexico's image in the world, serving as a defender of national identity and exercising its authority over domestic and foreign films alike. In concerning itself with social control, it acted as an arbiter between conflicting social forces in the domestic arena. The Interior Department and its censorship apparatus were administratively autonomous of the other major state institutions, operating without regard to the financial implications of their decisions.

After the establishment of the censor's office, a presidential decree created formal regulations in August and set up a rating system that divided films into four categories. State censors intervened occasionally to shape the actual messages that filmmakers delivered, even if this was largely an effort to mediate between contending social forces and their ideas. Recognizing the wide popularity of the medium, censors sought to preempt any possible social disturbance that might arise in the aftermath of Mexico's violent and tumultuous revolutionary experience earlier in the century. Authorities sometimes vacillated under the pressure they faced from both the left and the right, but they shared with these groups a common, paternalistic view of the Mexican public as needing cultural guidance.[72] The Catholic Church's League of Decency exerted a powerful influence on the censorship regime, and state policy aimed accordingly at distinguishing among films appropriate to certain age groups.[73] The state's ongoing concern with Mexico's image and reputation became apparent in 1944, when the head of censorship declared that he would not allow the exhibition of any film that denigrated the country, regardless of whether it was foreign- or domestically produced.[74] Thus, state policy in this period was promotive of *mexicanidad*.

While state cultural institutions enjoyed high parity with economic ones, the establishment of the Cinema Bank, with its more commercially oriented mandate, created a contradiction in the array of institutions overseeing the sector. The Bank's financial imperatives put film producers in the position of having to concern themselves primarily with the short-term commercial viability of their pictures, rather than enabling them to focus on film quality or the long-term strength of the business. This attracted entrepreneurs, but it dissuaded others from engaging in filmmaking. Consequently, the state devised a number of institutional solutions to this tension. While the privately owned Cinema Bank received original support from the National Bank of Mexico and the state development bank, it was detached from the larger structure of economic governance. As early as 1944, the head of censorship joined union representatives and a leading producer to form a tripartite commission to oversee film interests, expecting renewed international competition in the days ahead. In December 1947, a new National Film Commission was established with a mandate to "save the national cinema" and elevate the aesthetic quality of movies.[75]

Once the industry entered what would become a perpetual crisis in the late 1940s, state authorities sought to maintain its basic economic viability as a precondition to shaping the substantive messages it delivered. The Cinema Industry Law of 1949 is instructive in this regard. It contained numerous provisions linking the industry's waning economic strength to its deteriorating quality, calling for a resolution of its problems for the dual purpose of the cinema's economic development and moral elevation (Art. I).[76] Toward this end, the law provided for the awarding of prizes, the enhanced use of film in education and as a means of cultural diffusion, the imposition of screen-time quotas, the waging of a public relations campaign, and myriad other initiatives to promote high-quality films serving the national interest (Art. II). While the film law also stipulated that its violation could result in serious fines or even the arrest of violators, subsequent enforcement was exceedingly weak, and three years later, in 1952, the law had to be revised under a new plan.

State promotional efforts in this early period were rooted in the assumption that it was possible to return the industry to its wartime strength with the right combination of incentives, disregarding the unique circumstances that had facilitated its initial development and success in the 1930s and 1940s. While observers throughout the world in the first half of the century had high hopes and exaggerated fears regarding the social influence of the cinema, the particular political importance of the Mexican industry may have emanated from the coincidence of its early success with the period of state- and nation-building that followed Mexico's emergence from civil war. By rising to prominence at a historical moment when Mexico was first articulating its modern, twentieth-century self-conception and building accompanying state institutions, the film industry attained a unique place in the state's effort to define Mexico's postrevolutionary national identity.[77] Cultural policy would reflect that privileged position for the next two and a half decades, as state officials took for granted that the industry had to be supported to define, to defend, and to promote a certain conception of the country.

Another of the most enduring early institutional developments that solidified the state's cultural institutions appeared in the 1949 Film Law, which established a General Cinema Directorate (*Dirección General de Cinematografía*).[78] The Cinema Directorate was placed in the Department of Interior, given direct subsidies, and endowed with broad powers regarding film policy, including over matters of censorship. The 1949 law also provided for the establishment of a National Council of Cinematic Art, again created as a consultative body for the Department of Interior and intended partly to promote the cinema's "moral and artistic perfection." Presided over by an Interior official, the fourteen-member Council included representatives from a broad range of other departments, the Cinema Bank, all the industry's subsectors, and both unions.[79] When the 1949 law was amended in October 1952, the Interior Department's institutional preeminence was strengthened even further.[80]

Throughout the 1950s, the cultural ideals and artistic expectations of state elites confronted the rampant commercialism of the majority of investors in film, who saw the industry in strictly economic terms. Policy makers attempted to correct the conflicting incentives generated by the state's cinema institutions and return the industry to its former international status. With a rich artistic tradition on which to draw, and experience as a cinema leader in the developing world, such ambitious cultural goals seemed entirely attainable for Mexico. The state's film-related institutional apparatus was safely ensconced in the powerful Department of Interior, which dominated the decision-making process for all aspects of state film policy and had the full support of President Adolfo Ruiz Cortinez after 1953. Cinema institutions were therefore highly autonomous from the other departments, with the culturally privileged position of filmmaking in Mexico insulating it from the rest of the state apparatus. Since filmmaking was not subject to the same criteria applied to other fields, producers had easier access to commercial credits than most other industrial sectors in Mexico.

These institutional advantages notwithstanding, filmmakers were anything but oblivious to financial pressures. Regardless of the tenor of state rhetoric about the cultural importance of filmmaking to Mexican national identity and the country's image in the world, a commercial imperative very often triumphed over artistic goals. While the Cinema Bank's ostensible purpose was to guarantee long-term financial support for a beleaguered cultural industry, it deployed its resources in such a way as to assure the dominance of commercial tendencies and encourage a shortsighted approach on the part of filmmakers. The state-owned Cinema Bank provided filmmakers with generous and discounted production loans, but it expected timely repayment, with interest; and producers had to bear the full costs of any losses they generated.[81] Such an incentive structure punished risky, creative innovations and encouraged a closed circle of industry insiders to invest as little as possible, recycle past successes, and shun new talent. Consequently, the Cinema Bank failed from the outset to find a formula for cultivating simultaneously both commercial success and artistic achievement.

Despite growing global pressures and increasing competition from television, Mexican cultural policy remained largely promotive in this period. With the first regular television broadcasts beginning in July 1950, state policy makers witnessed this newer medium quickly replace the cinema as a favorite national pastime.[82] A stark contradiction began to appear between the stagnating film industry and continued assumptions about its positive role in identity-building. The early successes of the film industry added to its perception in governing circles as a crucial medium of national expression, resulting in state concern at the highest levels. Efforts to resuscitate the ailing industry were initiated at the top. In May 1951, President Miguel Alemán delivered a message to cinema workers, declaring filmmaking to be more than just an industry; it was an expression of what it meant to be Mexican. He described the cinema as a cultural and artistic medium that crossed borders so the world could understand better the country's history, social organization, folklore, and landscape, exhorting filmmakers to meet and exceed this high standard in order to maintain Mexico's positive image in the world.[83]

The following month, in June 1951, Undersecretary Ernesto P. Uruchurtu of the Ministry of Interior elaborated further on the perspective of his ministry, which had fullest responsibility for cultural policy toward the cinema. Uruchurtu described approvingly the power and influence of the Mexican cinema as a form of artistic expression and an educational medium, but he warned of the serious dangers associated with exporting a false image of Mexican national identity. He called on the industry to conform to moral and artistic norms, demanding that filmmakers convey "the truth" about Mexico as a united, optimistic country, proud of its past and committed to joining the world's most developed countries. Uruchurtu contended that the cinema was the best vehicle for communicating these ideas to the world, relating President Alemán's call to the sector's various branches to create not only a strong industry in the economic sense, but an organ of "cultural diffusion" and a means of national self-improvement. He noted that the president, as a former Minister of Interior, was especially interested in the film industry, which he affirmed could count on the permanent support of public authorities.[84]

While a promotive state cultural policy was in place in the 1950s, officials wrestled with the difficulty of buttressing the industry economically while cajoling it into improving the quality of its output. By August 1952, industrial decay had begun to yield qualitative decadence: The Mexican Academy of Cinematic Arts and Sciences, for the first time since its establishment in 1942, could not find a film worthy of the award for best film that year, in its equivalent of Hollywood's Academy Awards. To counter this decline in film quality, the state revised the cinema law in the ensuing weeks to strengthen its promotive stand on cultural policy. These changes included requiring that national films be exhibited for a period of at least 50 percent of the total screen time at any given theater and authorizing the building of new state-sponsored production studios. By intervening to promote production and assure what it considered to be appropriate and fair exhibition, the new regulations were designed to protect the public interest and shape national identity. Yet this policy could only be realized at a cost to certain segments of the industry, since exhibitors earned greater revenues from showing foreign films than from domestic ones.[85]

In the next few years, the Cinema Bank played a powerful executive role in promoting particular films that were claimed to serve the national interest.[86] Bank credits were nominally intended to support films with socially redeeming value. To this end, the Bank began promoting expensive, export-oriented superproductions filmed in color, as well as engaging in coproduction with Spain, Cuba, France, Italy and other producing countries.[87] Entertainment in the capital went through a period of austerity, as the growing middle class called increasingly for the removal of vice from public culture and state officials responded by stepping up censorship and shutting down some theaters. This development resulted in a puritanical phase in which half of all films in 1954 were melodramas involving middle-class professionals.

Despite the continuation of a promotive cultural policy, the overall quality of many Mexican films remained dismal, and symptomatically, Mexico's national film awards, the *Arieles*, were suspended in 1958 for nearly a decade and a half.[88] When the administration of Adolfo López Mateos bought both major television channels in 1960, many observers concluded that the state soon would nationalize the film sector to rescue filmmaking from what was perceived to be its fatal flaw: merciless commercialism.[89] While this did not happen, the state did take control of the Churubusco Azteca studios and the 365 theaters that constituted the Jenkins monopoly, forming the *Compañia Operadora de Teatros* (COTSA) in a move that increased its leverage in national exhibition circuits.

In November 1960, moreover, the Chamber of Deputies discussed new film legislation that had important implications for the "national interest." Under the proposed legislation, all foreign movies had to be shown in their original language, with subtitles and not dubbing, thereby reducing their popular appeal.[90] But the legislation also contained draconian revisions of censorship regulations that reflected the increasingly illiberal tenor of state cultural policy. It prohibited the distribution or exhibition of any film that reduced the moral level of spectators; attacked decency, peace, or public order; spurred crime or the mockery of justice; attacked the social institutions of marriage, family, respect for parents, and patriotism; or distorted history. Many of Mexico's leading literary and cultural figures signed a statement denouncing the draft law as an unconstitutional constraint on artistic freedom, and the law was tabled despite the financial benefits it promised the industry. To the dismay of many of these artistic figures, so they claimed, state authorities had come to misunderstand the purpose and potential value of cultural production. Nineteen sixty, after all, was the same year in which producers, state bureaucrats, and industry representatives selected a forgettable but profitable box-office hit, *La Cucaracha*, as Mexico's official entry at the Cannes Film Festival, rather than a film the industry had mocked: Luis Buñuel's grim but critically acclaimed critique of Christianity, *Nazarín*. When the official entry became a laughingstock of the competition, and *Nazarín* triumphed in unofficial screenings, the befuddled and angry Mexican delegation departed early.[91]

State cultural institutions remained highly autonomous from their economic counterparts throughout the remainder of the 1950s and into the 1960s. New film legislation in December 1960 would have allowed the Department of Industry and Trade to regulate film imports and exports, but the law never made it out of the Senate because of the controversy surrounding its severe censorship provisions. Under the López Mateo administration, the quality of filmmaking sank to new lows, as Luis Buñuel left to work in Europe, director and STPC leader Alejandro Galindo turned to making cheap melodramas, and Emilio Fernández found himself boycotted by producers.[92] The country registered substantial if uneven developmental gains and resulting social upheaval, though somehow none of this was reflected in the cinema. The state institutional structure, cinema-related and otherwise, was locked in place by a powerful alliance of vested interests that assured organizational longevity.

The state moved to reign in some of the industry's perceived excesses and to promote its own vision of Mexican culture. It did not stop financing private producers, though it found itself increasingly at odds with producers over issues of film content and quality.[93] In 1962, in responding to calls by producers for more assistance, the Minister of Interior emphasized, on behalf of the president, the need to improve film quality before more aid would be forthcoming. This was somewhat of a departure for the Mexican state, which in contrast to Argentina, did not make subsidies and monetary awards contingent on film quality.[94] Mexican producers therefore resisted making films like Buñuel's *Nazarín*, which had won its share of international laurels but had no box-office success.[95] Still, when the state announced its intention to restructure the film industry in August 1964, it stressed in the most dramatic terms the industry's importance to Mexican national life, its power as a medium of expression and education, and the unexpected reach of filmmaking as an ambassador of art, culture, and national thought.[96] In time, censorship became slightly less restrictive, though heavy political controls were kept in place for films made or written by critics of the government, such as the acclaimed writer Carlos Fuentes, especially after the events of 1968 began to tear at the Mexican social fabric.[97]

The Cinema Bank, for its part, continued to finance a large percentage of production, though it had little actual control over which films received public funding. The Bank's principal criterion for granting credit remained commercial, even if the justification for its existence was more broadly social.[98] This meant that when the industry found itself in distress again in the early 1960s, the Bank came into conflict with the producers and unions that were most directly in charge of failing production.[99] Seeking a solution to the sector's many problems, authorities took larger parts of the industry under direct state control without resorting to outright nationalization. The Cinema Bank announced its intention to restructure the industry in 1964, and a particularly effective head of the Cinema Directorate was appointed late in the year. Still, the Bank made no significant structural changes to alter its relationship with other film-related state institutions in the remainder of the decade.[100]

With filmmaking verging on collapse, President Luis Echeverría began restructuring the film industry immediately in January 1971. His administration initiated the most drastic period of institutional innovation and autonomy in the history of Mexican cinema, and these changes resulted in even greater institutional parity. When Echeverría took power, the state already owned the Cinema Bank, the COTSA exhibition circuit, and Churubusco Studios; it also had a share in the two major overseas distributors, *Peliculas Mexicanas* and *Cimex*. Over the course of the next few years, the state nationalized *América* studios, the *Cadena de Oro* exhibition chain, and the remaining large domestic distributor, *Peliculas Nacionales*. It also took control of *Procinemex*, a private company founded in 1968 to promote the film sector.[101]

State centralization of the production process accompanied a number of related institutional changes. In 1974, Cinema Bank director Rodolfo Echeverría established *Conacine*, the National Cinema Corporation, the first of three state production companies formed to handle the growing administrative burden of a rapidly expanding public sector. Two more production companies followed in 1975, *Conacite Uno* and *Conacite Dos.*[102] *Conacine* introduced changes in financing, whereby the state bank contributed 80 percent of the funding for a given film and film workers themselves put up the remaining 20 percent. After both sides recovered their investments, earnings were supposed to be split equally. This usually took two to three years to occur, however, and producers' long-standing practice of inflating budget proposals meant that the Bank actually contributed most of the funding.[103] The result of the state's effort at institutional reorganization was the perpetuation of a highly promotive policy, despite its economic costliness and the sometimes disappointing results.

The democratic opening in the cultural arena under Luis Echeverría after 1970 had policy implications for filmmaking, since this was an area of particular interest to the new president. Censorship did not disappear in this period, as family organizations demanded closer state oversight, but it was scaled back so that filmmakers calling for free expression could show movies not previously screened. The president's brother, Rodolfo, as head of the Cinema Bank, reestablished and presided over the defunct Academy of Cinematic Arts and Sciences to give out the *Arieles*. He also removed the artificially low ceiling on cinema ticket prices, refurbished low-level urban cinemas, and improved advertising for national films.[104] The Bank tried, as well, to encourage the development of a more auteurist cinema. To promote quality filmmaking, the state inaugurated in 1974 the National Cinematheque (*Cineteca Nacional*), as required by the 1949 film law but until then left unrealized. The following year the state also founded a film school directly across from the *Cineteca*.[105] The year 1976 represented a high point in the Mexican state's promotive involvement in the film industry. While the overall volume of production was declining, the thirty-six state-produced films of that year exceeded for the first time the number of privately made ones, as a new generation of well-trained filmmakers entered the field.[106]

Eventually, after six years of what amounted to the nationalization of the industry, a reverse transformation of the state's cinema institutions began after the José López Portillo administration arrived to usher in a period of low institutional parity. In 1977, López Portillo created a powerful new state agency to coordinate Mexico's growing array of mass media, the Directorate of Radio, Television, and Cinema (RTC), headed by his sister, Margarita, whose tenure was by most accounts disastrous for the film sector.[107] By placing the film sector under the auspices of an agency devoted to the electronic communications media, the new administration changed the institutional relationship between the Mexican cinema and the state. The RTC was placed in the Department of Interior, along with its affiliates, the General Cinema Directorate and the Cinema Bank. But it turned to encouraging traditional, commercially oriented private production, which had virtually collapsed under the previous administration.

This was a turning point in state institutional development. Reflecting a very different perspective on the appropriate role of the state in public culture, the RTC presided over the disestablishment of much, though not all, of the public sector film apparatus. One of the three state production companies, *Conacite Uno*, was liquidated immediately. Margarita López

Portillo closed the Cinema Bank, which stopped financing filmmaking in 1978 and was mothballed in 1979, amid what appeared to be trumped up charges of corruption. All of the Cinema Bank's administrative functions were reassigned to the RTC.[108] In closing the Bank, President López Portillo presided over the elimination of a state industry in which, as Secretary of the Treasury and Public Credit six years earlier, he had authorized investing one billion pesos.[109]

Just when the cinema's successes were beginning to produce larger audiences, the López Portillo administration initiated a change of course in Mexican cultural policy, beginning a long period of drastically reduced state involvement under a laissez-faire orientation. The prior administration's commitments forced the state to continue producing for a short period, but many of these films never were screened. The changed tenor of policy was evident in July 1979, when security personnel entered the state-owned Churubusco studios, arrested dozens for their opposition to the new policies, accused them of fraud, and imprisoned some for an extended period. As if to signal the final demise of a promotive cultural policy, the *Cineteca Nacional* burned to the ground in March 1982 because of the negligent handling of inflammable nitrates. In effect, rather than promoting cultural production, the state was sabotaging its own involvement in it.[110]

Under Miguel de la Madrid from 1982 to 1988, cultural policy was a distant second to dealing with national economic difficulties, beginning with the debt crisis. State authorities invoked strong budgetary constraints to avoid even minimal cultural promotion, as the administration gave pressing economic concerns full precedence over any expenditures on cultural policy.[111] The state wound down virtually all its major film-related commitments and liquidated most of its holdings in the industry, ceding the field to private actors to determine the direction of cultural production. The state seemed to have abandoned the idea of using film to promote a certain image of Mexico at home and abroad, or as a vehicle for expressing *mexicanidad*.

The creation of the RTC broke with the past by linking the cinema to television and radio, and this represented a significant step in realigning filmmaking with other, more strictly commercial enterprises in the mass media. This change accompanied the rise of Mexican television conglomerates, such as *Televisa*, which had established its own film production company, *Televicine*, a few years earlier in 1978. The cinema thereby lost its privileged institutional position, even more so when the Mexican economy began shifting toward a more private-sector and market-dominated orientation after the peso devaluation and debt crisis of 1982. This transformation accelerated after Mexico's accession to the General Agreement on Tariffs and Trade in 1986 and North American Free Trade

Agreement in 1992, both of which required further movement toward more liberal, internationally oriented policies.

Filmmakers themselves resented bitterly the policy change, as well as the degrading seediness of the new commercial cinema and state authorities' seeming indifference to this. Its clear short-term financial successes aside, the resurgent private production of the period was described by one Mexican critic as "industrially feeble, socially useless, culturally nil, and aesthetically impoverished."[112] After recovering from the shock of denationalization and its consequences, filmmakers became hopeful again in 1983, when the de la Madrid administration announced plans to establish a federal agency to coordinate film policy: the Mexican Cinema Institute (*Instituto Mexicano de Cinematografía*—IMCINE). IMCINE, however, was subordinated to the RTC, and it proved disappointingly ineffective in its early years.[113] The state also rebuilt the *Cineteca Nacional* in 1984, charging it with administering Mexico's film rating system. Its financial control was shifted in early 1987 from the defunct Cinema Bank to the National Bank of Public Works and Services (*Banco Nacional de Obras y Servicios*—BANOBRAS). While the stated purpose of the change, according to the president, was to maintain Mexican cinema as "a living testament of quality in national culture, and also as a source of employment," its effect was to diminish the organizationally autonomous standing of cinema.[114] The administration's priorities focused on coping with the severe national economic crisis, with little state attention remaining to deal with cultural concerns.

These developments notwithstanding, certain institutional changes regarding IMCINE eventually did appear to elevate, at least nominally, the place of cultural production in the Mexican state's decision-making structure. In early 1989, control over IMCINE was transferred from the Department of Interior to the Department of Public Education's newly constituted National Council for Culture and the Arts (*Consejo Nacional para la Cultura y las Artes—CONACULTA*). President de la Madrid created the latter in the 1980s as an independent and autonomous patron of the arts and culture, with its director appointed directly by the president. In the mid-1990s, discussions were held to elevate it bureaucratically to the status of a separate, cabinet-level department on par with the others, though with no eventual success.[115]

IMCINE in the 1990s retained a dual agenda that entailed promoting a commercially successful quality cinema. Toward this purpose, it attempted to coordinate international coproductions and private-sector investment, aid in the marketing and distribution of national productions, and organize retrospectives of Mexican films.[116] Much of the IMCINE personnel had worked previously for state producers and distributors, the Cinema Bank, or other segments of the industry, providing a measure of institutional memory and continuity.[117] Additionally, IMCINE controlled a relatively small pool of funding, the Support Fund for Quality Cinema (*Fondo de Fomento a la Calidad Cinematográfica*), which was established by the state in the 1980s as seed money for a select number of promising film projects.[118] It also had a measure of administrative authority over the state film school and Churubusco-Azteca studios. For their part, filmmakers hoped that IMCINE would acquire a standing like that of the well-regarded Institutes of Fine Arts and Anthropology, which also were placed under the bureaucratic authority of CONACULTA. The critical successes of some of the productions that IMCINE aided in the 1990s signaled to some observers the possibility of a revival in quality cinema.[119]

The downsized industry of the 1990s had neither substantial access to public investment, nor was it involved any longer in the promotion and development of Mexican national identity. Where the state did intervene, financial constraints were invoked to limit the direct availability of investment capital. State economic institutions dominated high-level decision making, such as when the Department of Commerce wrote new film-related legislation in 1992 to assure that the law would be in harmony with NAFTA requirements.[120] Persistent demand for inexpensive commercial films contributed to their continued production into the 1990s. The result was the financial failure of critically well-regarded art films alongside the commercial success of poor-quality, made-for-video production and television exports. Unlike in the past, however, state decision makers no longer had an institutionally supported interest in promoting a strong cultural policy, despite the economic constraints. As a result, those pockets of bureaucrats and filmmakers who sought state support for quality cinema no longer had their interests embedded in, or defended by, state institutions.

Nothing in subsequent developments would reverse the laissez-faire trend. The state rented out Churubusco-Azteca studios and other facilities to foreign filmmakers in the 1980s and 1990s, eventually moving to privatize them. While retaining minimal control over cultural production, the state appeared to be moving toward a more complete transfer of initiative to the private sector. The Department of Interior's General Cinema Directorate maintained the censorship authority (*Subdirección de Autorizaciones*), continued to grant import and export permissions, and was charged with registering producers and authorizing foreign companies to work domestically. Perhaps more importantly, state cultural policy in Mexico no longer treated the film sector as crucial to the Mexican state or nation. Cultural policy became predicated on the view that Mexico no longer could afford to support a fully developed local film industry, as it had for decades. Alternatively, the state assisted a select group of individual filmmakers in realizing higher quality productions, doing so largely through coproduction arrangements with other national industries requiring minimal state direction.

What is most striking about the long trajectory of Mexico's cultural policy is that it remained promotive for so long, indeed for much of the century, despite its economic costs and extended periods of industrial crisis. Even in the darkest days of low-quality industrial filmmaking, political leaders joined state authorities in articulating the merits of government support for what was considered a vital national industry. A subsequent period of laissez-faire cultural policy coincided with the liberal economic orientation of new administrations. But the retreat of the state from an activist and promotive cultural policy only followed the dismantling of state cultural institutions in the 1970s and 1980s. Mexico's laissez-faire cultural policy by the end of the 1990s was entirely unsurprising, given the reconfiguration of the state's institutional balance and the resulting newfound dominance of economic institutions touting liberal ideas about the most appropriate relationship between the state and cultural production.

Conclusions

Institutional parity has proved to be a useful way of understanding the privileging of one set of policy concerns over another in the Egyptian and Mexican contexts. Four general patterns emerge from a review of the evidence. First, institutional configurations in Egypt and Mexico reflected political and economic origins that often did not relate directly to concerns in the film sector. Institutions were created, changed, and sometimes abandoned for reasons that frequently had more to do with exogenous trends in politics or the economy than anything specific to filmmaking or even cultural production. In Mexico, for example, the Department of Interior was prominently involved in the film sector for decades. Yet this stemmed primarily from the Department's early development as a postrevolutionary instrument of social control in an era of popular instability. As an accident of history more than by conscious, long-term plan, state authorities deemed it appropriate to give the Interior Department bureaucratic oversight of moviegoing. Similarly, Egypt's Ministry of Culture was established in the latter part of the 1950s, but not initially as a mechanism to support the film sector or defend its interests. It came into being when the cultural functions of the Ministry of National Guidance—originally housed in the Ministry of Education—proved ill-fitting with the latter's institutional mission of controlling and disseminating information. Once created, all of these institutions had effects that did not necessarily reflect policymakers' conscious and intentional plans.

A second pattern relates to the association of a high degree of institutional parity with the heightened responsiveness of state policies to cultural interests, even at the expense of economic ones. When parity was high, states promoted filmmaking because of the power and autonomy of their official cultural institutions in relation to other state bodies. In Mexico in the early 1970s, for example, a newly empowered and independent Cinema Bank, led by a well-connected director, had the resources and bureaucratic authority to nationalize much of the film industry. The Bank took a number of decisive, financially costly steps to reform and advance filmmaking as an instrument of state cultural production that had been in crisis for over two decades. The subsequent formation of an even more powerful institutional rival in 1977, however, brought the state's cinema institutions under the authority of economic actors, and led to associated changes in cultural policy. In 1960s Egypt, similarly, the rising bureaucratic profile of the Ministry of Culture gave significant impetus to autonomous and independent cultural production. Freeing public filmmaking from economic constraints proved culturally invigorating, though it was financially ruinous in the long run. This institutional change did not guarantee the production of high-quality films, but it did allow for more experimentation and less deference to commercial success as the sole animating force in the industry. Under high institutional parity, states were willing to pay a premium for political control, even if this meant sacrificing to some degree the long-term economic well-being of the sector.

A third pattern in the evidence is the mirror image of the second: a low level of institutional parity all but assured that state policies were relatively unresponsive to cultural interests or laissez-faire in their handling of such concerns. Under these circumstances, the film sector was obligated to survive on its own merits as an economic enterprise, and commercial tendencies prevailed. Egypt in the 1970s, for example, saw the separation of the state cultural apparatus from what had become a more powerful Information Ministry, and even the temporary dismantling of the Ministry of Culture at the end of the decade. These changes left their mark on policymaking by stripping the industry of well-positioned bureaucratic allies, and the result was a wave of commercialism that culminated in the video boom of the 1980s and the industry's virtual disintegration in the 1990s. In Mexico of the 1980s and 1990s, calls went out for renewed state aid to the cinema; yet, new institutional arrangements simply reinforced the rising dominance of a particularly overt form of commercialism. Low institutional parity defined film as a commercial product, meaning that state leaders were less likely to expend resources on promoting a cultural revival in filmmaking.

A final pattern lies in the extent to which state institutions did not always attain their purported goals. Indeed, even when high institutional parity led to promotive cultural policies, the quality of cinema and the health of film industries often deteriorated. This may reflect the extent to which even strongly held state interests can fall prey to the dysfunctionalism of weak bureaucratic implementation. For example, Mexico's Cinema Bank was established in the late 1940s with a broadly cultural and social purpose, but it came to represent the pursuit of financial success in filmmaking. It served the industry faithfully in one sense through the steady provision of credit, but undermined it in another by offering credit only to those producers who had the capacity to repay their loans quickly and efficiently. While institutional parity structures policy choices, it does not account for the particular manner in which policy is implemented, or for the policy's broader, long-term consequences.

The early successes of both the Egyptian and Mexican film industries were made possible in part by the unusual political and economic circumstances of World War II in each of their regions. Wartime expansion may have given an entire generation of Egyptian and Mexican filmmakers false hope that they could maintain large-scale, economically sound, commercially oriented national film industries in perpetuity. In Egypt, the industry's success in spreading the country's cultural influence in the Middle East also colored expectations, especially on the part of state policy makers seeking the extension of Egyptian influence throughout the entire multistate domain of the Arab nation. Filmmaking in Mexico became bound up in issues of domestic national identity and nation-building, as state authorities in the Department of Interior sought to defend and cultivate Mexico's image in the world.

These similarities aside, Egypt and Mexico responded in diverse ways to similar global pressures over time. While Egyptian cultural policy shifted from laissez-faire to promotive and back primarily to laissez-faire, Mexican policy remained promotive for much of the century until the mid-1970s. These differences may have been inspired partly by domestic regime changes and economic crises. But the persistence of cultural policies in both countries, even after changes in the domestic conditions supporting them, points to the enduring role of state institutions. The relationship between cultural and economic institutions helps to explain changing support for policies that were either culturally rich but economically costly, or economically rational but culturally impoverished.

Part III

National Responses to Globalization

6

Conclusions and Implications

Robert Gilpin has described the inevitable tension that exists between state policies of "Keynes at home" and "Smith abroad." In national economic policy making, states sometimes intervene in domestic industry to achieve socially desirable or politically motivated ends, while advocating free trade internationally to advance the interests of firms that profit from open exchange. This same antagonism plays out in the domain of cultural industries like filmmaking, except the tension is between a state's cultural and economic goals, rather than its domestic and foreign ones. The political choices that result from the policy-making process reflect the weight of competing economic and cultural interests, articulated through existing domestic institutions to produce complex and sometimes contradictory outcomes. Reconciling the conflicting imperatives of the state's varied interests is an especially acute problem that manifests itself explicitly in the governance of the film industry.[1]

Under the right constellation of circumstances, the push and pull of economic and cultural interests work at cross-purposes, as one segment of the industry, such as film exhibitors, advocates a state policy of liberal disengagement, while another, perhaps film producers, lobbies for protectionist involvement. The resulting tension between cultural and economic interests is especially prominent when their inconsistency is a long-standing feature of the policy-making landscape. The extent of the disparity between economic and cultural goals is a measure of the likely difficulties state authorities will encounter in sustaining policies. This dynamic can be doubly problematic in a globalizing world, where any contradictions between commerce and culture are thrown into sharp relief, much as tensions between domestic and foreign economic policies are revealed under the right circumstances.

This book has explored these tensions by making an argument about the consequences of national variation in the economic structures and state institutions that mediate international competition in film. It has asked and answered a series of related questions: Why do states vary in their responses to globalization? Why do some states accept economic losses in exchange for cultural gains? Who bears the costs of cultural protectionism and who reaps the benefits? In the following, I review and extend the study's principal responses to these questions, as well as exploring more fully the relationship between economic and cultural globalization. I then draw out the implications of the argument by looking at several related issues and problems in political economy.

Principal Findings

Trade and Cultural Policy

The book's departure point has been a distinction between the economic and cultural aspects of policy making, rooted in the assumption that this distinction is essential to understanding how global competition affects a seeming oxymoron, the cultural industries. The distinction is artificial, or constructed, since cultural policy sometimes has economic components and economic policy has cultural implications. Nonetheless, any account of state responses to globalization may require a large degree of parsing. This is particularly important for a long-standing and multifaceted cultural industry like filmmaking, since its involvement in general processes of change in both the international economy and world culture subject it to a diverse array of potential influences. In the cases examined, it was evident that trade and cultural policy were not driven only by economic variables, such as the desire of powerful blocs of firms for relief from the competition meted out by foreign imports. Policy choices also were affected by domestic institutional arrangements, which shaped the capacities and interests of actors.

Trade and cultural policies sometimes were complementary, or mutually reinforcing, and at other times, they contradicted each other, not embodying a single, consistent tendency. Such a pattern was evident, for example, when a liberal trade policy toward distributing and exhibiting film imports coincided with a promotive cultural policy, such as active state involvement in production (e.g., Mexico 1960s). Alternatively, the state sometimes restricted the activities of foreign firms in the domestic film market, while taking a more skeptical view of film production as a cultural interest worthy of financial support (e.g., Egypt 1980s). As the disparate elements in state bureaucracies worked against each other, this made it essential to look below the surface of formal policy pronouncements to see the contours of the ongoing policy-making battles occurring within the state.

International Competition

Competition alone does not explain all the variation in policy outcomes, either cross-nationally or historically, though the globalization process is not without significant effects. In some cases, mounting competitive pressures sharpened the contradictions and tensions in the policy-making arena. Such pressures had different effects on the various subsectors of the industry and therefore on eventual policy choices. Film producers nearly always felt international pressures most acutely, except in the relatively unusual instances of coproduction with foreign companies. Distributors were either helped or harmed by rising competition, depending on whether they traded in foreign films or competed with them. Exhibitors sometimes were uniquely positioned to benefit from globalization, particularly when they relied on foreign films for a large share of their receipts. All three subsectors worked consistently to defend their interests by lobbying for their policy preferences. Yet their success or failure depended less on the direct strength of international competition and more on the specific structural and organizational characteristics of the domestic industry at a given historical moment.

Despite its uneven impact on policy outcomes, international competition did strengthen the position of consumers. Especially after the globalization of the industry intensified in the late 1960s, commercial imperatives grew even stronger. This change contributed to rising consumer power in a globalizing world, where increasingly seamless and unmediated demand drove the supply of cultural products to consumers endowed with more choices and discretionary buying power than ever before. In the absence of active promotion by the state, large-scale noncommercial filmmaking generally lacked a continued capacity to operate, since it required either a captive domestic audience that had few alternatives or state support to underwrite production costs that could not be amortized by large receipts. In some cases, a successful commercial branch of the industry could provide an indirect subsidy to its noncommercial counterpart. In the United States, documentary and personalized filmmaking reaped tremendous benefits from the pool of talent, technology, financing, and infrastructure created and supported by Hollywood. In most instances outside the United States, however, noncommercial filmmaking faced increasing financial constraints driven by the growing integration of the industry worldwide.

These findings also underscore the analytical significance of distinguishing between the economic prerequisites of filmmaking, which have long exerted a powerful influence on the industry, and the changing nature of popular demand for cultural products like film. The demand side of the equation—audiences—grew increasingly important throughout the century and serves as the key factor linking the economic and cultural dimensions of globalization. Even monopolistic firms were obliged to be attentive to audiences because the elasticity of demand for luxury goods enabled moviegoers to forego the cinema if they grew dissatisfied. Firms in competitive markets, likewise, could not afford to ignore audience preferences because of audiences' capacity to choose among competitors. New technologies since the invention of television have added to this necessity, with the growing availability of leisure-time alternatives that include the rapidly expanding domain of the Internet. In this sense, individual consumer demand connected the cultural dimension of film commodities to their economic requirements in a globalizing world.

As a result, the cultural content of filmmaking under globalization has become the subject of renewed controversy. On the one hand, increasing capital mobility for film financing, growing corporate control over movie distribution pipelines, and fuller market integration all would appear to require investment only in the mass-market oriented, homogeneous film products that are the hallmark of cultural globalization. Increasing capital mobility in what remains a particularly risky sector, for example, would seem to encourage the pursuit of large markets that require lowest-common-denominator products. Rising corporate dominance of global distribution since the late 1960s would appear to result in the exclusive distribution of "one-size-fits-all" movies that are selected by the leading distributors. Growing market integration might seem to yield a single, homogenized film product that can be profitable everywhere.

On the other hand, evidence from the film sector suggests that the competitive pressures of economic globalization have not led to a uniform, homogenized state of cultural globalization. American-based capital has been relatively mobile in the film sector since the 1950s, but it has been powerless to control film content in locales with access to funds from culturally diverse sources. In Egypt, for example, films that are funded by foreign investment from Saudi Arabia or France tend to differ considerably from U.S.-financed imports. The growing strength of large film and entertainment multinationals has been counterbalanced by the growth of niche markets and the fragmentation of consumers into demographically narrow, if geographically far-flung, audiences. Finally, the integration of the world film market as a whole was well under way by the 1930s, with no obvious associated effect on the diversity of filmmaking throughout the world. While the economic pressures of globalization have been a reality for decades, cultural globalization remains far more limited in scope than the conventional wisdom claims.

Market Structure

Domestic market structure provides a persuasive, if partial, explanation for trade policy outcomes in the film sector. By looking at the number of firms, the ease of entry into the sector for new firms, and the number of consumers, it becomes apparent that variations in market structure affected policy outcomes. Markets affected policies by shaping the political capacity of the sector. Firms in competitively structured industries found it more difficult to cooperate politically to secure their policy preferences. Competitive markets usually resulted in liberal policies, which generally benefited exhibitors and could not be prevented by producers. Such industries often sank into a pattern of subsectoral conflict that undermined their political influence as a whole, even if particular subsectors fared well at times. In contrast, firms in monopolistic markets were better positioned to cooperate politically to win their preferred policies. When faced with strong international competition, industries with this kind of market structure sought and obtained state protection, which benefited all firms that were part of the industry. Strong interfirm connections across the subsectors mitigated conflict and aided in the presentation of a united political front, preventing defection by those firms that might have profited from more liberal policies.

A related finding, moreover, is that subsectoral politics had significant effects on the content and aesthetic quality of film production. This mattered greatly because it influenced the popularity of these industries' products with their audiences, therefore affecting their long-term economic viability. While economic cooperation between firms (i.e., monopoly) helped them to secure protectionist policies, it also sometimes impoverished film production from a qualitative standpoint by eliminating the competitive mechanism. Badly made pictures weakened the industry's standing with audiences over the long run, producing a vicious circle of demands for protection that only made such protection more necessary thereafter. Even in the relatively rare instances in which monopolies had the power and political influence to limit severely the availability of foreign imports as substitutes for their own lower-quality local production, the high elasticity of demand for moviegoing meant that audiences could simply stop going to the cinema. In a number of instances in the case studies, this is precisely what happened and with devastating consequences for the industries. To make matters worse for monopolists, technological innovations like television created entertainment alternatives that rendered their market power and exclusivity irrelevant from the vantage point of most audiences.

Private-sector monopolies, in particular, sheltered film industries from the participation of new talent and had little purpose other than the defense of their own economic interests. In the best of times, when international competition was weakened by exogenous factors like war, private-sector monopolies operated relatively successfully in financial terms, just as Hollywood rose to prominence during the studio era of the 1920s and 1930s. While Hollywood's monopoly was ended by the application of U.S. antitrust laws, remaining private-sector monopolies elsewhere typically grew rigid, stagnant, more fully closed to newcomers, and eventually lost their audiences, as happened in Mexico beginning in the late 1940s. State-owned monopolies in both Egypt and Mexico were somewhat exceptional in the aesthetic contributions of their best work, since they were animated partly by a social purpose that aspired at least nominally to higher-quality filmmaking. Both the Egyptian and Mexican state monopolies, however, proved untenable in the long run, as the shifting political landscape eroded their support and mounting financial expenditures made it impossible for them to remain self-sufficient.

If monopoly power sometimes undermined the commercial value of production, competitive markets were found to enhance it, though only under very particular circumstances. Economic conflict between firms (i.e., competition) led almost inevitably to political failure in pursuing policy preferences, but it enriched cultural production by forcing filmmakers to be more responsive to the wishes of audiences and critics. This is not to say that competitive markets always produced the best-quality filmmaking. Competition often bred a commercialization that led to conservative pandering to the most popular fads. Yet from the perspective of the industry's long-term economic health, and the mitigation of a need for protection that came with monopoly-induced stagnation, competition usually was good for the industry as an economic enterprise. Commercial success was a double-edged sword. It created otherwise nonexistent opportunities for filmmaking by supporting the ongoing development of the industry and its members. Simultaneously, commercialism reinforced the role of financial considerations in all aspects of the undertaking.

A consistent logic prevailed, then, in terms of the consequences of market structure for the cultural content of film industries. Competitive markets contributed to the qualitative improvement of production from an economic standpoint, while monopoly led to its eventual degradation under most circumstances. Over time, competitive markets therefore caused political weakness, commercial strength, and liberal policies; monopolistic markets caused political strength, commercial weakness, and protectionist policies. The aesthetic judgments inherent in the latter findings are broad generalizations at best. They are valid from the standpoint of the likely commercial success of total productive output with large popular audiences, rather than being subjective critical evaluations of aesthetic merit. Competition forced filmmakers to give audiences what they wanted, an outcome with its own, sometimes unfortunate, consequences for creativity in the industry. Monopoly freed filmmakers from such constraints, but often with poor results all around.

On a parallel, international level, the same mechanism that protected domestic firms from global competition—weakly developed domestic markets—also undermined the viability of these firms as healthy and successful participants in the international film trade. In a variety of cases, the underdevelopment of local markets had crosscutting effects: It de-linked the domestic industry from the threat of foreign competition, while preventing national filmmakers from tapping into the potential economic gains from expanded exports abroad. As in other, unrelated industries, film producers who were capable of protecting their markets from foreign competition lost the incentive to improve their output in the eyes of domestic and foreign audiences alike. By advocating an authoritative rather than market-based allocation of resources, these firms were successful in defending their immediate positions. But this success came at a high long-term price in a globally integrating industry. Such firms took few compensatory actions to develop local markets and exposure to international competition in the form of globalization pressures often resulted in their retreat even further.

Institutional Parity

Finally, the degree of institutional parity has notable effects on cultural policy. Under conditions of high parity, state cultural institutions are autonomous from economic ones. The overarching structure of institutions functions more like a market than a hierarchy, and cultural interests have strong representation. When cultural policy is made under such conditions, state cultural institutions seek to be the principal locus of decision making, and the state takes a more activist role in promoting national culture. Under conditions of low parity, state cultural institutions are subordinate to economic ones. The structure of institutions functions more like a hierarchy than a market, and cultural interests are more weakly represented. Policy choices are implemented by bureaucrats using economic yardsticks to measure success, and the state regulates accordingly. In either case, institutional change is difficult, because interests coalesce around the alternative structures: Ministry of Culture officials never willingly renounce their raisons d'être and seldom relinquish authority to other bureaucratic entities.

By this logic, not only do the institutions themselves matter, but the bureaucratic relationship between contending institutions matters, too, especially when they have overlapping claims of authority. The institutional balance reflects historical struggles over the scope and domain of legitimate state authority. It has a decisive effect because it determines the very basis on which policy choices are made, and it creates conditions for the perpetuation of one set of interests over all others. State cultural institutions link decision makers to the promotion of a particular conception of national identity in film. The strength or weakness of such links has clear consequences for the role of the film sector in national life. This is true even in liberal states, where, in the absence of state cultural institutions, few administrative mechanisms exist to regulate production and to support state intervention in filmmaking.

The degree of institutional parity serves most fundamentally to affect the exact nature of the state's interest in the film sector. Under low parity, economic criteria are paramount in decision making because economic institutions have full authority over the film sector. The social purpose of these institutions, observable in the rules that they make and play by, identifies filmmaking as a commercial activity, pure and simple. Alternatively, under high parity, a different purpose pervades those state bodies that have the authority to make and implement cultural policy. In such cases, economic efficiency is sacrificed for the benefit of a state-defined cultural interest, such as the promotion of a certain idea of national identity or the articulation of specific national values. In either case, decision makers are constrained by the state interest identified and embedded in the institutional arrangements that they inherit from past choices. Through the intervention of political leaders, that interest can evolve and be transformed over time, but only in tandem with suitable institutional changes. Cultural policy proves to be surprisingly persistent and costly to change.

By defining the exact nature of the state's interest in filmmaking, the institutional structure also has important consequences for the application of domestic or international considerations in policy making. Institutional factors influence whether film is considered by state authorities to be principally a tradable commodity under the domain of foreign economic policy, or a cultural commodity implicated mainly in domestic concerns. The United States, for example, has never engaged in substantial domestic regulation of filmmaking, even after the Paramount decrees of the 1940s. At the same time, it has promoted film exports more actively than any other state in the world. This contrasts with the actions of many other states, which display much more domestic involvement in the industry but much less international concern. Differences in state institutional arrangements—and therefore in the articulation of state interests—helps to explain this distinction, and an awareness of such variation is especially vital when states interact in trade negotiations.

A related if unsurprising irony is that the creation or strengthening of an autonomous state regulatory institution, such as a Ministry of Culture, does not guarantee commercial or critical success for a particular national film industry. High institutional parity does lead to policy changes and stronger conscious efforts to promote filmmaking, but state institutions alone cannot bring audiences into theaters or win the praise of critics. Low parity, however, usually is associated with the dominance of commercial tendencies, whether economically successful or not. Without a formal institutional mechanism and a set of incentives for rewarding high-quality, noncommercial film production, filmmakers generally must respond to the preferences of mass audiences. If film industries become commercialized, they remain so unless state authorities intervene to insist on substantive or qualitative changes in production.

Dynamic Interaction

Interactions between the variables sometimes produced crosscutting effects, with consequences for the success and long-term viability of any particular national industry (see Table 6.1). When high parity accompanied monopoly, this generally led to both protectionist economic policies

Table 6.1 Long-Term Effects of Markets and State Institutions

		Market Structure	
		Monopolistic	Competitive
Institutional Parity	High	Economic Decline & Cultural Vitality	Economic Success & Cultural Vitality
	Low	Economic & Cultural Decline	Economic Success & Cultural Decline

and promotive cultural policies. Such a combination often assured the industry's cultural vitality for a time, but it also had the potential to cause industrial stagnation and entrenched rent-seeking. Eventually, film production declined in cases where state bureaucrats worked hand in hand with industry insiders to milk the system for their own benefit.

Its polar opposite—low parity and competitive markets—assured the supremacy of economic incentives in the industry. The unique pathology of this combination, however, was the complete commercialization of filmmaking, with little room for other kinds of production. If filmmakers failed in the face of domestic or international competition, eventually they ceased to operate. The mixed outcome of high parity and competition produced more favorable results, as industries profited from state assistance but remained compelled to survive on their own merits. The most unfortunate combination entailed low parity and monopoly, which provided neither state support nor the economic incentives of competition, precipitating the industry's overall decline. In the end, whether through benign neglect or active support, state policies have been instrumental to the success—and failure, in many cases—of the industry.

Cultural Autonomy versus Consumer Sovereignty

As a whole, these findings suggest a vital trade-off that has not been explored sufficiently in globalization debates: The desire for both cultural autonomy and what might be called consumer sovereignty. In its starkest manifestation, an unmistakable tension exists between the objectives of some state authorities and the interests of individual moviegoers in a globalizing world of integrating markets. State authorities attempting to shape national identity via cultural policy require strong and autonomous institutions to do so, even when their motives reflect only the pursuit of short-term political interest. These state institutions are inherently constraining of personal autonomy, empowering authorities to make choices and take actions that affect the domestic availability of cultural commodities. In contrast, some individual film consumers seek a full range of cultural choices in accordance with the personalized norm of consumer sovereignty that is emerging in market economies. Sovereign consumers aspire to choose everything for themselves, rejecting state involvement as elitist and antidemocratic. Inevitably, the freedom of choice that consumers have is inversely related to the extent of the constraints imposed by state cultural policy: greater consumer choice requires lesser state intervention in defense of national cultural autonomy, just as limited consumer choice enables greater state involvement in defining national culture.

The success and failure of states in maintaining cultural autonomy in the world today depend on the extent to which the norm of consumer sovereignty has been embedded in domestic institutional structures. Consumer sovereignty represents a rejection of the legitimacy of state authority in the cultural domain. Where individuals have attained, through the democratic process or any other means, a prioritization of individual choice over state regulation, weaker national cultural autonomy and coherence is inevitable under globalization. In these cases, a strong Ministry of Culture, for example, is untenable. Alternatively, where institutionalized political support exists for limiting consumer choice, greater national autonomy and coherence is possible. Under such circumstances, a Ministry of Culture may act to define and preserve national culture by regulating and sometimes limiting the menu of choices available to citizens or subjects. Under either scenario, the ultimate arbiter of cultural autonomy is not the strength of international markets, but the political choices and state structures that continue to govern policy making. In this sense, cultural autonomy remains a matter of fundamental political process more than economic imperative.

The specific nature and value of state cultural policies varies greatly, since a strong cultural policy may be either parochial and politically motivated or more genuinely oriented toward shaping a coherent national identity and defending national cultural autonomy from undue external influence. On the one hand, state authorities may wish to regulate cultural discourse simply to control a potential arena of political opposition, assuring that no political space exists in which domestic rivals can articulate alternative conceptions of national identity and purpose. On the other hand, authorities may seek to preserve the autonomy of national culture, seeing powerful foreign states as genuine cultural and ideological threats. As the fulcrum of decision making, state authorities must play both domestic and international games simultaneously, keeping one eye on local political rivals and the other on more global concerns.

The idea of a closed and exclusionary national identity is antithetical to liberal internationalist notions of personal freedom. Yet, perhaps counterintuitively, the norm of consumer sovereignty is built on an expectation of diversity and freedom of choice that may not be fully realizable without some degree of cultural control. In cases where individuals expect to be able to choose from both locally made, nationally resonant films and the global products of filmmaking multinationals, the market dominance of the latter may have the capacity to undermine the making of the former. An economically weak local film industry cannot compete easily with its international rivals, and thereby may require some measure of state protection by authorities seeking, at a minimum, to use the language of national culture to advance their agendas. In this sense, the practical limits of what amounts to cultural populism, as expressed by the liberal ideal, are found in the economic realities of cultural production in a world of national states.

Broader Implications

To what extent are findings from the film trade generalizable and relevant to larger concerns? The three major divisions of filmmaking have analogous segments in other industries: film production is similar in some ways to any basic manufacturing activity; film distribution is comparable to the marketing of a whole range of products; and film exhibition is equivalent to retail sales. The concepts of market structure and institutional parity therefore may be applicable to industries facing strong global pressures and potentially subject to noneconomic criteria in state decision making. Certain aspects of the politics of the film sector do not differ in any appreciable way from those of many other industries. In nearly any industry, for example, the structure of markets may either precipitate conflict or facilitate cooperation between firms. The resulting nature of industrial relations affects outcomes in the larger political field, with an associated impact on policy making for the industry. For this reason, if the extent of economic conflict and cooperation embodied by market structure is important in constraining or facilitating political action in the film sector, it may also do so in other, similarly organized industries. This is particularly true for those in which market structure varies greatly and powerful international competitors affect the relationships between domestic actors.

Yet cultural production like filmmaking does differ from other industries. Film is different because it has a latent potential to be treated as such by state authorities. This potential is a result of the mixed, even schizophrenic historical development of the motion picture as both a medium of entertainment and a mode of cultural expression. Under certain circumstances, when the state actually has a clearly articulated cultural interest in filmmaking, policy makers differentiate film from more conventional commodities, subjecting it to noneconomic criteria in the policy process. At other times, when no such interest has been defined, film is more comparable to other commodities produced by the service sector. As a result, what is important is not that any given film actually has cultural significance, but that state authorities sometimes believe it does, and treat it accordingly. By exploring the conditions under which this occurs, this study may contribute to extending the classic Weberian tension between economic and cultural factors to international political economy.

Public and Private Interests

State authorities construct artificial definitional boundaries between the appropriate domains of public and private economic activity, with the distinction between the two being a function of historical contingency more than short-term political choice. Decision makers sometimes, but not always, identify the film sector as having public cultural significance, somewhat akin to that of a public good. In such cases, authorities distinguish filmmaking from other kinds of economic activity by regulating it on the basis of noneconomic criteria, giving film a place in the public sphere. There is wide cross-national variation in the scope of the public interest attributed to all cultural products like film. State decision makers vary in their handling of filmmaking because the institutions shaping their choices differ significantly in the criteria by which state policy is made. This variation is important, since it affects the degree and form of state involvement, which has a range of implications for the public interest.

In more liberal polities, guardianship of the public interest in filmmaking and other cultural industries is delegated to private economic actors. The public interest is equated with the sum of individual actions, and the preferences of individual consumers determine the nature and availability of cultural products. In such an environment, state authorities attempt to assure property rights and provide the basic conditions needed to carry out economic activity, but they involve themselves only minimally in regulating the substance of such activity. Where it is dominant, the liberal model even holds with mass media like television and radio broadcasting, though the latter ostensibly are more closely regulated. In the United States, for example, television and radio broadcasters are required by law to serve the public interest as a condition of being granted a license from the Federal Communications Commission for the right to operate on the limited number of available frequencies. Yet television and radio content is driven almost entirely by consumer demand, and the market mechanism acts as a means of allocating resources to firms that serve a public interest defined largely by popular taste. Public television and National Public Radio are the exceptions that make the rule.[2]

In contrast, where an alternative schema of values is institutionalized in an interventionist state, the public interest is protected more directly by state authorities. Such authorities define more expansively the appropriate sphere of legitimate state involvement in cultural production, usually to include the ideational and ideological elements associated with national culture and identity. Decision makers regulate both the economic and legal conditions supporting film and other forms of cultural production, as well as the content of the products themselves. This regulation may take the form of a range of interventions, from censorship and limitations on free expression to subsidies and quotas. This suggests that, in the area of cultural production, cross-national variation in the size and scope of the public domain hinges on the evolution of the state's institutional apparatus. The expansion and contraction of the public domain is an ongoing but uneven process, subject to contestation and fully implicated in the historical construction of state regulatory institutions.

Regardless of the nature of state institutional arrangements, filmmaking may potentially become involved in public life because of the particular attributes of the product, both cultural and economic. All films have the public goods characteristic of nonrivalness, even when private economic actors create them. For this reason, an important dichotomy of interest exists in film industries everywhere. Ongoing and successful film production requires at least some degree of responsiveness to the economic prerequisites of financial solvency. The film product cannot be made, circulated, and seen by mass audiences without a substantial investment on the part of some party. At the same time, the substance of the film itself sometimes is seen as having consequences for a larger public interest beyond the narrow financial interests of its makers. In a sense, the film commodity begins its physical existence like any other purely private economic good, made by an investment of capital and labor that generates returns and requires repayment. Yet once any film enters the public domain, it is transformed into an item of broader social significance that state authorities in some cases are inclined to regulate for its content.

In a world of sovereign states, the commercial and cultural aspects of filmmaking have contending logics. The economic characteristics of the film commodity make it highly tradable, and require of it the widest possible circulation to paying audiences. Its cultural characteristics, in contrast, give it a potential political significance, and therefore call forth efforts to control the medium by state authorities concerned about the effects of its substantive content. When the economic and cultural attributes are brought together in different combinations, they yield varied results. Since this tension is present in an increasing number of goods and services in the world today, analysts can expect further political conflict, in which locally produced cultural commodities are drawn into broader processes of global change. Since these commodities have locally derived meanings that cannot be transposed easily onto the larger, international arena without losing something, many may not survive such processes.

Only some cultural production lends itself readily to public presentation, but substantial cross-national variation exists in the public space accorded to artistic endeavors like filmmaking, painting, theater, or musical performance. The public role of the arts in Mexico, beginning at least with muralists like Diego Rivera early in the twentieth century, is quite striking, and its role in national politics is therefore not surprising. Very early on, a paternalistic state took an interest in, and virtually appropriated, the public space for the arts. While the Egyptian state has been equally paternalistic for at least two centuries, it has taken relatively less interest in deploying art and culture for political purposes. That said, the Egyptian state has a long history of varied degrees of involvement in cultural production: the city of Cairo's architectural wonders reflect the variation of state involvement and patronage of the arts, and one can map out the rising and falling fortunes of the state and its various regimes by the quality and sophistication of public art and architecture.

States and Institutions

The notion of institutional parity developed here has wider applicability to state institutions in areas other than cultural policy. States have very different regulatory paradigms for the many issues over which they extend their authority, and the attributes of state institutions governing these areas are potentially significant for structuring the relations between the state and social actors. Without a reasonably independent and authoritative state institution, dedicated to a specialized policy area or issue, it is less likely that policy makers will define and defend an interest in such an area in the face of social opposition. Without some form of institutionalized representation, state interests will not be articulated in the rough-and-tumble world of bureaucratic interplay or in the regulation of social action. This is true for both older issues, such as national culture, and for newer areas into which state authority has been extended more recently, such as the environment and human rights. In short, the degree of institutional parity affects whether these areas are deemed appropriate for state intervention, or better left to the care of private actors.

An example from another area illustrates the point. As Peter Haas once noted in a study of environmental epistemic communities across states bordering the Mediterranean, the political success of any given group of experts depended on the existence of a Ministry of Environment.[3] He found that the Mediterranean basin's southern littoral states of Egypt and Algeria were more successful in making and implementing policy reforms than their northern counterparts, France and Italy. This was because Egypt and Algeria had specialized and relatively autonomous state institutions dedicated to the environment, which afforded local epistemic communities access to the policy-making process. In other words, the degree of institutional parity mattered, as measured in this case by the autonomy of their Ministries of Environment from other actors in the state apparatus. Italy, on the other hand, had no such institution, and therefore no state agency existed to defend environmental interests. France's state environmental institution, Haas also notes, was subordinated bureaucratically to the authority and interests of the Ministry of Foreign Affairs. With low institutional parity, the structure of state decision making gave other actors in the bureaucracy a capacity to trump environmental interests and affect policy decisively.

If institutional variation has consequences for state action and interests, then the diversity of these interests must be scrutinized and traced back to the accidents and idiosyncrasies of institutional creation, rather than being taken for granted. The path dependency of institutional development during state formation belies the notion of a single, universal, inevitable, and necessary set of policy-making agencies that all states must have.[4] Some states have a Ministry of Culture or a Ministry of Environment, while others do not; the causes of such differences are not so mysterious as to be immune to comparative examination. The persistence of even the most dysfunctional and ineffective state institutions speaks to their ongoing significance. History does not necessarily erase all signs of failed institutions, which often continue to have ongoing if diminished influence.[5] For this reason, the historical origins of variables like institutional parity should be explored more systematically.

Under specific and definable conditions, economic efficiency may be sacrificed for other values that are sufficiently important in a given institutional context. For the substantive issue under consideration, state authorities may accept economic losses in favor of cultural gains, and they may do so as a basic matter of institutionally mandated political choice. This occurs when the state institutional apparatus empowers specific officials to choose costly ways of furthering the public interest. International competition may render such choices relatively expensive by providing what would be efficient alternatives and by punishing states economically if they choose to ignore them. Yet nothing about the process of globalization eliminates the sovereign authority of state decision makers to choose economic inefficiency. Even in a globalizing environment, it is commonplace for rational decision makers to choose economically inefficient alternatives to meet compelling political requirements that are shaped in some measure by the structure of the state institutional apparatus. Nothing inherent in globalization removes the contingency or complexity of institutional variation that is a legacy of the diverse historical experiences of state formation throughout the world, making such choices tenable in the first place.

This viewpoint stands in sharp contrast with a popular analytical perspective in which efficiency reigns over, if not rules, all state decision making. In a globalizing world, where efficiency is considered the "Ur value," according to Robert Dore, this study challenges the permanent, inherent primacy of economic efficiency as the sole interest to which state policy makers will devote themselves.[6] The findings detail the particular circumstances under which decision makers are willing to forgo economic considerations for other kinds of ends. Sacrificing commerce for culture is just one example of the trade-offs that are not only possible, but likely in a world of institutional diversity. In another policy-making area, nothing prevents states, for example, from pursuing less aggressive developmental paths to safeguard the environment, though the specific structure of institutional authority may affect policy choices in vital ways, possibly even frustrating the wishes of the dominant social coalition for a time.

These conclusions are consistent with the conception of institutions found in historical and sociological approaches in the literature. This is clear from the evidence presented relating to where state cultural institutions came from, what they did, and how they changed. These institutions generally originated from well-defined political circumstances, more than as attempts to solve problems of collective action, coordination, or other challenges. Their specific characteristics were strongly influenced by the available alternatives that served as models for institution building. Institutional forms also were shaped by random and unexpected historical contingencies, such as the outbreak of World War II, and exogenous technological developments like the rise of competing media. In regulating film, state institutions were active in ways that included defining the limits of the state interest in culture, structuring decision makers' policy choices, legitimating state action, providing criteria by which to allocate public resources, and helping authorities to understand outcomes. In so doing, they encapsulated the values of dominant social actors, reflecting earlier political choices and privileging the powerful.

The study also is consistent with an historical institutionalist perspective on political and economic change. The latter generally holds a discontinuous view of change and sees continuity as the norm, showing how institutional forms explain the stubborn persistence of state policies. Such an historical approach expects change to be rare, disruptive, and anomalous, therefore requiring explanation. More narrowly rationalist approaches, in contrast, are based on a continuous and adaptive conception of change, whereby lengthy periods of stasis demand explanation.

In the cases examined, institutional change did not occur incrementally, as adaptive responses to gradually evolving circumstances like the strengthening or weakening of film markets. The general array of state institutions overseeing the film sector tended to remain intact for extended periods of time, in some instances long after becoming dysfunctional from the standpoint of its original architects. When state institutions finally did change, this often occurred quickly, but never without political cost and effort, as state authorities moved to transform a mechanism that had stopped serving the political purposes for which it was originally created. This explains the otherwise puzzling delay between changes in political leadership and changes in cultural policy, as well as the inconsistency evident in mixed policy outcomes, such as when political leaders opted for a promotive cultural policy but a liberal trade policy.

Most noteworthy is the relative incapacity of state authorities to change the institutional structure. If this were not the case, institutions always would be secondary to the fleeting interests and objectives of individual political figures, and they would be changed much more often than is the case. Institutional landscapes persist, along with the kinds of solutions to political problems they make possible. Egypt's Ministry of Culture, for example, was legislated out of existence in 1979, only to have its functions and personnel transferred to a newly formed Supreme Council for Culture. The Ministry itself was formally resurrected several years later by officials who acknowledged its utility in serving a political purpose for which no other state institution existed. The Ministry offered a ready-made bureaucratic capacity to engage nonmilitarily what would become a sustained challenge from the Islamist opposition over the nature of the Egyptian state and national identity. Likewise, in the Mexican context, the persistence of the Cinema Bank for nearly forty years was testimony to the difficulty of achieving institutional change in the face of the entrenched interests that came to rely on it so heavily. Despite its gross inefficiencies and net drain on state coffers, only with the sweeping liberalization measures of a determined and reform-minded de la Madrid administration did it see its final demise.

Similarly, the *absence* of an American institutional capacity in the cultural domain is a comparable example of the persistence of institutional landscapes. In national discussions of Hollywood and the purportedly deleterious effects of its products on American youth, the absence of any real institutionalized mechanism with which to shape, much less control, cultural production has reduced policy makers to making empty threats about regulation while imploring the industry to improve its product and regulate itself better. Not surprisingly, no recent public figure has suggested creating an entirely new cabinet-level post concerned with national cultural production to complement the economic focus of the American regulatory structure in this area. This is at least partly because of the constraints imposed by the ideological dominance of market-oriented solutions to collective social problems in the United States. But accompanying the dominance of market thinking is an institutional structure that empowers state policy makers only in the economic domain, leaving them impotent in the cultural one. Changing institutions is possible, but it is politically costly and difficult.

Markets and Culture

What are the implications of the above for understanding the complex relationship between market-based economies and cultural production? One likely issue is the importance of considering the broader social effects of entirely market-driven cultural production. In the United States, and under liberal capitalism in general, the buying power of the middle class throughout the twentieth century has assured that producing and selling entertainment has been a lucrative business enterprise. But many cultural critics see American cinema as simple entertainment: a temporary diversion that enables workers to relax, forget their concerns momentarily, and recharge their energies. Even if shorn of its Marxist ideological veneer—capitalist filmmaking as social opiate—this depiction of American cinema acknowledges the frivolousness attributed to filmmaking by some cultural critics. In other places and at other times, however, including in the United States, the cinema is and has been closely linked to efforts at public education. When created under certain circumstances, film is defined by the state as a public good, contributing to intellectual and ethical development, as well as to the public-mindedness of citizens, albeit sometimes in an overbearing and paternalistic way.

Hollywood's critics on the right characterize the cinema today as a public "bad," contributing to the putative moral crisis of a generation of young people. They and other critics of commercial cinema often work from a simplistic understanding of the cognitive influence of movies and the social psychology of audience members, even if media theorists have failed to communicate the complexity of how audiences read, interact with, and assimilate cultural production. Nonetheless, market-driven cultural production does have a pathology that has become more prominent in recent years in the most advanced capitalist economies. By reinforcing the social identity of moviegoers as consumers more than citizens, a growing portion of commercial cinema in the age of globalization may be of dubious social value. The global "victory" of the market late in the twentieth century is not without a darker side, since market-driven cultural production does not always yield socially optimal results. Even Adam Smith concluded that unfettered markets would not produce everything needed by society, including in the areas of popular entertainment and education, where state encouragement might be necessary to protect the public good.[7]

The changing relationship between markets and cultural production must be understood in the historical context of the development of market economies. In *The Great Transformation*, economic historian Karl Polanyi described the commodification of labor as a necessary prerequisite to the growth of an efficient self-regulating market economy. For this to occur, large numbers of individuals had to be removed from their social structures and transformed into discrete, mobile, denatured, anonymous units of labor. Similarly, the commodification of culture—its complete subjugation to the laws of supply and demand—is the inevitable consequence of a fully market-driven system of cultural production. This occurs when all cultural products are made for individual purchase and consumption. In its most extreme form, electronic commerce on the World Wide Web represents this emerging trend in the world economy. It is characterized by the most efficient of all possible economic transactions: exchanges that are conducted between atomized, anonymous, hyperindividuated consumers who are stripped of all identity, and online dot-com businesses that do not occupy finite places in the physical world.

Such a transformation of the social place of cultural production began to occur in twentieth-century advanced capitalism. It followed a comparable, parallel change in the nature of social cleavages in much of the contemporary world. The old Marxist class-analytic focus was well suited to social identities in the nineteenth century, when most people were industrial or agricultural producers of one kind or another. But it may be less relevant to the patterns of identity emerging in recent years, as consumer identities become more salient than the relationship of particular classes to the means of production. The commodification of culture articulates well with the rise of consumerism in a globalizing world, since it represents a process of fragmentation in the repertoire of choices available to individuals at any given moment. Cultural commodities are the archetypal items of consumption in such a world, constituting a further development of the connectedness of the modern economy.

Just as Polanyi focused on the unavoidable costs inherent in the transformation to a market economy, the potential social costs of capitalism are evident in the film domain today. Polanyi was not a Luddite; he neither regretted the great transformation, with the opportunities it created nor called for a return to the premarket past. But he did warn of the cost of this kind of change, and he paid particular attention to those who paid that cost. A loss of cultural diversity is not an inevitable by-product of market-led cultural production; yet, it remains a distinct possibility. Such a danger is similar to environmental threats to biological diversity, since cultural cross-fertilization has enriched even the most powerful and dynamic societies throughout history. Hollywood has attracted the talents of a rich variety of foreign filmmakers over the course of the past several decades, and these individuals have brought new perspectives that could not have emanated from southern California.

The winnowing of the cultural field would lead to the permanent disappearance of some of today's sources of social variation. This possibility comes in conjunction with the grinding down of social beings, their detachment from social roles, and the atomization of individuals as both discrete units of labor and, most importantly, consumption. This could occur in the culture trade with the wrenching of cultural production from its local contexts, as manifested in the seemingly innocent cultural pastiche created when people are free to pick and choose from a broad palette of cultural products available to them today. If cultural diversity in the world matters, global market competition alone may not be the way to maintain, let alone encourage it.

Regime Type and Globalization

The findings also underscore the complex relationship between regime type and state intervention in cultural production: even under similar regime types, policy may vary remarkably, just as different types of regimes may share comparable policies. Trade protectionism and promotive cultural policies occur wherever market structures and state institutions create the appropriate incentives and conditions for state intervention, regardless of regime type. Liberal democratic regimes tend to involve themselves only minimally in the film sector, while nondemocracies are more likely to intervene. Yet this is less a function of democratic governance per se and more a product of the aversion to state intervention that is inherent, by definition, in liberal institutional arrangements. Both democratic France and authoritarian Brazil in the 1970s, for example, were heavily involved in their respective film industries, even if the democratic transition in Brazil eventually prompted institutional reforms that eliminated financial support for the state film agency. Elsewhere, a number of democratic regimes experienced an ongoing oscillation in the degree of state intervention in the sector, as in India and Britain. Finally, some authoritarian regimes all but ignored their film sectors, only taking notice after institutional developments defined a state interest in cultural products like film.

All of this points to a dilemma of cultural production. Strong state involvement in promoting the film industry and other cultural activities often entails an attempt to control the field of cultural expression and repress the articulation of alternative subnational identities. This occurs most frequently under the guise of efforts to build or protect national identity. Such efforts are always coercive and exclusive in some sense, as the state attempts to impose a particular vision of nationhood on all the inhabitants within the territory. In a sense, truly national cinemas are constructed at a cost to other, subnational identities, regardless of the nature of the ruling regime. For this reason, no simple relationship exists between regime type and the policy objectives pursued by the state. Just as international market pressures produce indeterminate results in the realm of state policy making, the indeterminacy extends to the impact of regime type on the substance of filmmaking.

In the cases examined, state intervention in cultural production often led to nefarious forms of censorship, which were designed to limit the expression of contrary ideas or conceptions of identity. This reality notwithstanding, the efforts of filmmakers to circumvent state coercion and cultural restrictions sometimes cultivated the artful deployment of imagery, allusion, and other literary techniques that eluded state censors and resulted in more powerful filmmaking. Authoritarian regimes did not automatically destroy cultural creativity, just as democracy was no guarantee of it and could not necessarily inoculate film industries against cultural or ideological excesses. For regimes in transition, political liberalization from authoritarianism did not ensure the revival of high-quality filmmaking any more than a deterioration of democracy undermined the capacity of the industry. Surely, the most repressive regimes usually stifled cinematic creativity, but state efforts to co-opt and provide patronage for cultural production sometimes yielded a surprising sophistication that belies the commonplace image of shoddy, transparent propaganda and warns of the state's capacity to manipulate images and ideas.[8]

Globalization and the Culture Trade

Globalization has become the most prominent metaphor for changes occurring in the post–cold war world. It is a highly contested term, generally viewed favorably by those close to the centers of world political and economic power, and with suspicion by those at greater distance. This division holds true both domestically and internationally. Developing countries are analogous to the domestic poor under advanced industrial capitalism in their social location in the world order. Their limited capacity to shape the culture trade is clear. Wealthy countries are equivalent to the domestic rich in their buying power and the influence this gives them in the international culture trade. Globalization creates winners and losers in cultural production and trade, as in any other area, but the political power of individuals, social groups, and states determines who will join the ranks of those benefiting and those losing in the process.

Public discussion of globalization has reflected two very distinctive perspectives, each rooted in different assumptions about the nature and consequences of the phenomenon. Domestic and foreign critics alike have depicted it in ominous, threatening terms, while expressing a fear of its potential impact. Globalization, for these observers, portends a loss of national control in several areas, including cultural expression and diversity. The latter are said to be threatened by the leveling and homogenizing effects of powerful, impersonal, and irresistible economic forces. Globalization is depicted as an only slightly modified form of imperialism, with these observers using some of the same rhetoric to describe it. Critics of cultural globalization have joined environmentalists, human rights activists, and labor activists to express these concerns, directing their anger most vocally at powerful international actors and institutions, such as the World Trade Organization, the International Monetary Fund, and the World Bank. Their shared concerns notwithstanding, criticism of globalization is weakest in the cultural sphere, since environmental, human rights, and labor activists tend to make the most dramatic and readily grasped claims of potential harm to their areas of interest.

The leading alternative viewpoint takes its inspiration primarily from mainstream liberal economics. It holds that competitive international pressures are forcing state policy makers to adopt similar market structures, institutions, and policies. The coercive nature of the process aside, this perspective embodies a much more sanguine and benevolent understanding of globalization. It claims that rising interconnections in the world are creating irresistible incentives for states to liberalize their economies and polities and converge their cultures, driven by the ascendance of liberal economic principles in the post–cold war era. Under these circumstances, globalization is deemed to be transformative: It has unavoidable consequences for the independence of state decision making, the continued authority of policy makers over national economies, and the extent of cultural independence enjoyed by all countries. While globalization pressures are said to undermine the power of state authorities, an optimistic assessment of long-term prospects sees this change as benign or even beneficial, since it will give power directly to individuals and weaken the grip of state authorities on their lives.

This book challenges both the above perspectives. It does so most basically by demonstrating the extent of cross-national variation in state responses to international pressures and by arguing that national markets and institutions are more impervious to globalization than much of the conventional wisdom suggests. Markets and state institutions are like prisms, mediating international competition in ways that require their inclusion in calculations of the pace and direction of change. In other words, what is important is not just the fact of global competitive pressure, itself a long-standing international reality. This pressure has different effects on various parts of the industry, with distributional consequences for related social groups. Most fundamentally, observers need to acknowledge better the ways in which international competition becomes enmeshed in local political struggles. It does so through the formation of business alliances with local partners, by harming some subsectors and helping others, and by the direct effects of its goods and services on local markets.

In this sense, many globalization critics misrepresent their grievances to some extent. They attribute too much power and an almost willful malevolence to policy makers, assuming that state authorities are wholly unconstrained in their choices while disregarding the structural biases that preclude decision makers from supporting the choices that critics favor. Their perspective is overly voluntarist in its assumption of the power of elite actors to devise what amount to virtual conspiracies. In taking this approach, critics actually may underestimate the seriousness of the challenges they face. The achievement of their policy goals is not only a question of political will, since state authorities must also operate within the confines of institutional structures that limit their capacity to make changes, and that may contribute to the definition of state interests in ways that favor globalization. Critics stand in opposition to a broad historical process that is not likely to lend itself to reversal as a matter of normal politics, even in the face of opposition from an array of social forces.

That said, many globalization defenders misunderstand the nature of the process. They declare too readily the powerlessness of states facing international pressures and use the inexorable aspect of globalization as an excuse for what amount to socially regressive policies and business actions. In agent-structure terms, they hold an overly rigid and deterministic viewpoint that exonerates state authorities of responsibility for their political choices. This entails the reification of globalization and the self-serving declaration that it compels states and firms to make choices that they otherwise would not favor, ranging from trade liberalization to company downsizing. They underestimate the negative consequences of powerful international processes for the most poorly positioned groups and overestimate the ease with which such actors can make adjustments to assure their survival. The defenders even misjudge the difficulty of mitigating the worst effects of globalization, playing down the challenges for those struggling to meet them.

Despite the resilience of filmmaking and other culture-producing industries to globalization pressures, it is possible that some national film industries may fall by the wayside in the long run. National culture producers may opt out of international competition altogether, as they come under increasing *political* pressure to change their market structures and institutional arrangements to conform to the standards established by more powerful states. Some may accept the dominance of leading culture producers like the United States, which has been able to parley the seeming inevitability of globalization, and its ostensible benefits, into a demand that smaller producers accept policy reforms. The latter may even strive to make institutional changes that eventually allow them to begin treating filmmaking in accordance with standard economic assumptions about product homogeneity. Others, however, are unlikely to do so. France's resistance to American cultural domination is deeply entrenched in a set of sufficiently powerful and autonomous state institutions that are likely to persist. By promoting a conception of the rationality and universality of French culture, France's long-standing state institutions will continue to provide a counterhegemonic model of resistance for other countries to adopt and emulate.

The cultural dimension of the world order is less conspicuous to those Americans who take for granted the naturalness of U.S. dominance in world culture and who associate it with American democratic ideals.[9] Its construction and ongoing reproduction on behalf of the most powerful states is obscured by the common American claim that film is merely another entertainment commodity. Consequently, U.S. trade negotiators are prone to insisting on rapid and dramatic institutional reforms to bring states into line with the requirements of broader trade agreements, such as those contained in North American Free Trade Agreement or World Trade Organization provisions. They are likely to triumph whenever their negotiating partners have cultural institutions that are subordinated to economic ones and when globalization pressures put in serious danger the economic interests that these institutions promote. This is not to say that the political leadership of any given state is powerless in adjudicating between contending state interests. Yet state authorities' perspectives and definitions of what is important are likely to be influenced heavily by the prior arrangements and commitments of the state.

Ultimately, the most powerful and transformative globalization pressures are fundamentally political in nature. They may be cloaked, however, in the language of economic necessity. As costly cultural industries are targeted for criticism on the basis of their economic weakness, critics make the leap from pointing out their inefficiency to asserting their inherent inviability. Critics also invoke consumer choice as the principal normative criterion on which to base policy decisions, arguing, for example, that free trade in cultural products allows people to decide for themselves what kind of audiovisual products they wish to have. Framing the issue solely in economic terms privileges the most powerful, economically dominant producers, while espousing the disingenuous assertion that opponents are somehow elitist or undemocratic. It also allows policy makers to construe their choices as a defense of consumer interests and a responsible effort to confront the difficult economic realities of the era. The latter is especially persuasive wherever states are struggling with genuine developmental difficulties, and where cultural industries can be portrayed as a frivolous luxury that some states cannot afford.

Some technological changes may reduce, in fact, the capacity of states to regulate certain kinds of cultural flows. With the growth of video-on-demand over the Internet and direct television via satellite, world cultural flows are becoming more seamless and elusive of state control. This does not translate into a democratization of the sources of cultural production, since the spread of U.S.-made computing hardware has occurred in conjunction with the growth of American cultural "software" for use with such equipment. Other technological developments, such as digitization, may help to reduce filmmaking costs and democratize the opportunities to participate in cultural production, while at the same time undermining "national" or state-centric film industries. The combination of political pressures and technological developments may induce states to withdraw from traditionally significant areas of cultural production, such as filmmaking, in favor of other forms that seem more easily controlled and cost effective, such as television and video production.

State involvement in cultural production has been retreating in a number of places in the post–cold war world, with the rising influence of liberal ideology and calls by powerful countries like the United States for minimal state intervention. But the findings of this study run contrary to official American claims of the naturalness, necessity, and inevitability of nonintervention in cultural production. The latter perspective rests on liberal assumptions that, in today's world, place individual consumption at the heart of political economy. Such an attenuated view of social identity, however, is belied by the perspective that people are more than just consumers. They are citizens and subjects, with social and political relationships that exceed, in scope and purpose, the narrow confines of consumptive behavior. As long as the international system continues to comprise a collection of discrete sovereign entities, decision makers in some places will be empowered to intervene on behalf of, or even despite, the wishes and interests of their citizenry. Without eliminating the institutional diversity that marks the world of sovereign states, some places will have Ministries of Culture with recognized political mandates to intervene in the constant interplay between commerce and culture.

In many ways, the globalization debate is a replay of long-standing discussions of the relative weight of agency and structure in social causality. Proponents of globalization focus on the compelling nature of international structural forces. Critics, in turn, highlight the role of individual states as agents responding to global pressures. Yet both sides miss the far-reaching ways in which national structures themselves can vary and shape policy choices. To modify Marx's famous claim about history, states make their own policies, but not under circumstances of their choosing. Persisting variation in the domestic politics of sovereign states cannot be ignored. All governing authorities remain deeply rooted in the local political structures by which they seek to maintain themselves. Supporters of globalization can challenge, perhaps even replace, such national structures, but they will have to contend with them.

Notes

Chapter 1

1. As Sadat wrote regarding his effort to enter the barracks, "What would be the meaning of my struggle—of the very man I call myself—if I were to be reduced to a spectator when my *raison d'être* was taking shape ?" See Anwar el-Sadat, *In Search of Identity* (New York: Harper and Row, 1977), 105–7; and Jehan Sadat, *A Woman of Egypt* (New York: Simon and Schuster, 1987), 125–27. I thank Jehan Sadat for her recollection of the film title. Personal communication, November 1, 2006.
2. A provocative early version of the popular account is Kenichi Ohmae, *The Borderless World: Power and Strategy in the Interlinked Economy* (New York: Harper Business, 1990).
3. Thomas Guback, *The International Film Industry: Western Europe and America since 1945* (Bloomington: Indiana University Press, 1969).
4. Another work that combines international, state, and societal variables is Stephan Haggard, *Pathways from the Periphery: The Politics of Growth in the Newly Industrializing Countries* (Ithaca, NY: Cornell University Press, 1990). A valuable collection on business-state relations in the developing world is Sylvia Maxfield and Ben Ross Schneider, eds., *Business and the State in Developing Countries* (Ithaca, NY: Cornell University Press, 1997). Two other attempts to theorize state-society relations are Peter Evans, *Embedded Autonomy: States and Industrial Transformation* (Princeton, NJ: Princeton University Press, 1995); and Joel S. Migdal, Atul Kohli, and Vivienne Shue, eds., *State Power and Social Forces: Domination and Transformation in the Third World* (Cambridge: Cambridge University Press, 1994).
5. Prominent examples of the substantial literature in political economy linking domestic and international politics include Robert Keohane and Helen Milner, eds., *Internationalization and Domestic Politics* (Cambridge: Cambridge University Press, 1996); and Suzanne Berger and Ronald Dore, eds., *National Diversity and Global Capitalism* (Ithaca, NY: Cornell University Press, 1996).
6. An early review of globalization trends in the film industry is found in U.S. Department of Commerce, National Telecommunications and Information Administration, Special Publication 93-290, *Globalization of the Mass Media* (Washington, DC: Department of Commerce, January 1993). A comparative discussion of other sectors is found in the Organization for Economic Cooperation and Development publication, *Globalisation of Industry: Overview*

and Sector Reports (Paris: OECD, 1996). For a dissertation on the French film industry's response to globalization, written from a neo-Gramscian perspective, see Martine Danan, "From Nationalism to Globalization: France's Challenges to Hollywood's Hegemony," (PhD diss., Michigan Technological University, 1994). For a cautionary note regarding the extent to which globalization is occurring, see Robert Wade, "Globalization and Its Limits: Reports of the Death of the National Economy Are Greatly Exaggerated," in Berger and Dore, eds., *National Diversity and Global Capitalism*, 60–88.

7. Works emphasizing changes in global culture include that of leading cultural critic Fredric Jameson, *The Geopolitical Aesthetic: Cinema and Space in the World System* (Bloomington: Indiana University Press, 1992); and Fredric Jameson and Masao Miyoshi, eds., *The Cultures of Globalization* (Durham, NC: Duke University Press, 1998). See also David Harvey, *The Condition of Postmodernity: An Enquiry into the Origins of Cultural Change* (Cambridge: Basil Blackwell, 1989); Arjun Appadurai, *Modernity at Large: Cultural Dimensions of Globalization* (Minneapolis: University of Minnesota Press, 1996); Rob Wilson and Wimal Dissanayake, eds., *Global/Local: Cultural Production and the Transnational Imaginary* (Durham, NC: Duke University Press, 1996); Peter Golding and Graham Murdock, "Culture, Communications, and Political Economy," in *Mass Media and Society*, 2nd ed., ed. James Curran and Michael Gurevitch, 15–32 (London: Arnold, 1996); and Annabelle Sreberny-Mohammadi, "The Global and the Local in International Communications," in Curran and Gurevitch, eds., *Mass Media and Society*, 118–38; Malcolm Bradbury, "What Was Post-Modernism? The Arts in and after the Cold War," *International Affairs* 71 (October 1995): 763–74; and Ann Cvetkovich and Douglas Kellner, eds., *Articulating the Global and the Local: Globalization and Cultural Studies* (Boulder, CO: Westview, 1997).
8. This point is clearest when filmmaking is compared with more conventional industries—automaking, pharmaceuticals, electronics—or any of the other narrative forms of artistic and literary expression—painting, musical composition, or literature. Filmmaking shares something with both kinds of activity, giving credence to the proprietary claims of each side. For a discussion of semiotics and related matters in film, see James Monaco, *How to Read a Film: The World of Movies, Media, and Multimedia*, 3rd ed. (New York: Oxford University Press, 2000).
9. Like other services—haircuts or investment advice, for example—consumers assume some of this risk, since they must pay for admission to a movie before having any certain idea of its value to them.
10. For more on the GATS and TRIPs, see Bernard M. Hoekman and Michel M. Kostecki, *The Political Economy of the World Trading System: The WTO and Beyond*, 2nd ed. (New York: Oxford University Press, 2001).
11. Financial movements, not surprisingly, have outpaced labor integration in this sector, just as cultural flows have been much more rapid than the diffusion of production-related technology and information.

12. An important analysis of the trade-related issues in the culture industries is Patricia M. Goff, "Invisible Borders: Economic Liberalization and National Identity," *International Studies Quarterly* 44, no. 4 (December 2000): 533–62. Suzanne Berger and Peter Gourevitch both note the potential for trade conflict over domestic structures like cultural institutions; see Berger and Dore, eds., *National Diversity and Global Capitalism*, 16, 259. For more on Euro-American conflict in this area, see Annemoon van Hemel, Hans Mommaas, and Cas Smithuijsen, eds., *Trading Culture: GATT, European Cultural Policies and the Transatlantic Market* (Amsterdam: Boekman Foundation, 1996). A discussion of the French-American dispute in GATT is found in Toby Miller, "The Crime of Monsieur Lang: GATT, the Screen, and the New International Division of Cultural Labor," in *Film Policy: International, National, and Regional Perspectives*, ed. Albert Moran, 72–84 (London: Routledge, 1996). A review of changing French policy in the 1980s and early 1990s is in Susan Hayward, "State, Culture and the Cinema: Jack Lang's Strategies for the French Film Industry, 1981–93," *Screen* 34, no. 4 (Winter 1993): 380–92.
13. I use the term Hollywood interchangeably with the American film industry, particularly its commercially oriented major firms. Excellent studies of Hollywood's place in the American economy include Aida A. Hozic, *Hollyworld: Space, Power, and Fantasy in the American Economy* (Ithaca, NY: Cornell University Press, 2001); and Allen Scott, *On Hollywood: The Place, The Industry* (Princeton, NJ: Princeton University Press, 2004). Valuable histories of U.S. involvement in the world film trade include Guback, *The International Film Industry*; John T. Trumpbour, *Selling Hollywood to the World: U.S. and European Struggles for Mastery of the Global Film Industry, 1920–1950* (Cambridge: Cambridge University Press, 2002); Ian Jarvie, *Hollywood's Overseas Campaign: The North Atlantic Movie Trade, 1920–1950* (Cambridge: Cambridge University Press, 1992); and Kristin Thompson, *Exporting Entertainment: America in the World Film Market, 1907–34* (London: British Film Institute, 1985). The economics of the industry are presented lucidly in Harold L. Vogel, *Entertainment Industry Economics: A Guide for Financial Analysis*, 6th ed. (Cambridge: Cambridge University Press, 2004), and Colin Hoskins, Stuart McFadyen, and Adam Finn, *Global Television and Film: An Introduction to the Economics of the Business* (Oxford: Clarendon, 1997). A history linking economic factors to stylistic ones is David Bordwell, Janet Staiger, and Kristin Thompson, *The Classical Hollywood Cinema: Film Style and Mode of Production to 1960* (New York: Columbia University Press, 1985). Other useful works with an American focus include Tino Balio, ed., *The American Film Industry*, rev. ed. (Madison: University of Wisconsin Press, 1985); Paul Kerr, ed., *The Hollywood Film Industry* (London: Routledge, 1986); Gorham Kindem, *The American Movie Industry: The Business of Motion Pictures* (Carbondale: Southern Illinois University Press, 1982); Barry R. Litman, *The Motion Picture Mega-Industry* (London: Allyn and Bacon, 1998); David F. Prindle, *Risky Business: The Political Economy of Hollywood* (Boulder, CO: Westview, 1993); and Jason E. Squire, ed., *The Movie Business Book*, 2nd ed. (New York: Simon and Schuster, 1992).

14. This literature began with Charles Kindleberger, *The World in Depression, 1929–39* (Berkeley: University of California Press, 1973); and Stephen Krasner, "State Power and the Structure of International Trade," *World Politics* 28 (April 1976): 317–48.
15. See Thompson, *Exporting Entertainment*, 1985, and Thomas Guback, "Shaping the Film Business in Postwar Germany: The Role of the U.S. Film Industry and the U.S. State," in Kerr, ed., *The Hollywood Film Industry*, 245–75.
16. See especially Trumpbour, *Selling Hollywood to the World*.
17. Such a perspective is implicit in Steven S. Wildman and Stephen E. Siwek, *International Trade in Films and Television Programs* (Washington, DC: American Enterprise Institute, 1988); and Eli M. Noam and Joel C. Millonzi, eds., *The International Market in Film and Television Programs* (Norwood, NJ: Ablex, 1993).
18. Obviously there are multiple currents of liberal thinking on free trade in general; see Douglas A. Irwin, *Against the Tide: An Intellectual History of Free Trade* (Princeton, NJ: Princeton University Press, 1996).
19. For a relevant example, see Jorge Schnitman, *Film Industries in Latin America: Dependency and Development* (Norwood, NJ: Ablex Publishing, 1984), 3. The latter draws on the dependency theory of Fernando Henrique Cardoso and Enzo Faletto, *Dependency and Development in Latin America* (Berkeley: University of California Press, 1979); and Peter Evans, *Dependent Development: The Alliance of Multinational, State, and Local Capital in Brazil* (Princeton, NJ: Princeton University Press, 1979). Related theorizing is found in Immanuel Wallerstein, *The Modern World-System* (New York: Academic Press, 1974); and Immanuel Wallerstein, "Culture as the Ideological Battleground of the Modern World-System," in *Global Culture: Nationalism, Globalization, and Modernity*, ed. Mike Featherstone, 31–55 (London: Sage, 1990). A thorough study of the relationship between the U.S. film industry and banking is found in Janet Wasko, *Movies and Money: Financing the American Film Industry* (Norwood, NJ: Ablex, 1982). Influential discussions of cultural imperialism include Armand Mattelart, *Multinational Corporations and the Control of Culture: The Ideological Apparatuses of Imperialism* (Atlantic Highlands, NJ: Humanities Press, 1979); and Herbert I. Schiller, *Communication and Cultural Domination* (White Plains, NY: International Arts and Sciences Press, 1976). For discussions of cultural imperialism, see John Tomlinson, *Cultural Imperialism: A Critical Introduction* (Baltimore: Johns Hopkins University Press, 1991); and Tamar Liebes and Elihu Katz, *The Export of Meaning: Cross-Cultural Readings of Dallas* (New York: Oxford University Press, 1990).
20. A founding statement of the Third Cinema movement was made by Argentine filmmakers, Fernando Solanas and Octavio Getino, "Towards a Third Cinema," reprinted in *Twenty-five Years of the New Latin American Cinema*, ed. Michael Chanan, 17–27 (London: British Film Institute, 1983). See also Jim Pines and Paul Willemen, eds., *Questions of Third Cinema* (London: British Film Institute, 1989); and Roy Armes, *Third World Film Making and the West* (Berkeley: University of California Press, 1987).

21. For more on this, see Robert Vitalis, "American Ambassador in Technicolor and Cinemascope: Hollywood and Revolution on the Nile," in *Mass Mediations: New Approaches to Popular Culture in the Middle East and Beyond*, ed. Walter Armbrust, 269–91 (Berkeley: University of California Press, 2000).

Chapter 2

1. An essential discussion of comparative institutional contexts is Peter A. Hall and David Suskice, eds., *Varieties of Capitalism: The Institutional Foundations of Comparative Advantage* (New York: Oxford University Press, 2001). See also Suzanne Berger and Ronald Dore, eds., *National Diversity and Global Capitalism* (Ithaca, NY: Cornell University Press, 1996); and Herbert Kitschelt, Peter Lange, Gary Marks, and John D. Stephens, eds., *Continuity and Change in Contemporary Capitalism* (Cambridge: Cambridge University Press, 1999).
2. In this sense, filmmaking is similar to book publishing, a form of cultural production with very early links to capitalist development in Europe. See Benedict Anderson, *Imagined Communities: Reflections on the Origins and Spread of Nationalism*, rev. ed. (New York: Verso, 1991), 37–38. On the place of books in European cultural policy, see Annabelle Littoz-Monnet, "The European Politics of Book Pricing," *West European Politics* 28, no. 1 (January 2005): 159–81.
3. Peter Gourevitch, "The Second Image Reversed: The International Sources of Domestic Politics," *International Organization* 32, no. 4 (Autumn 1978): 881–912.
4. Jeffry Frieden, *Debt, Development, and Democracy: Modern Political Economy and Latin America, 1965–1985* (Princeton, NJ: Princeton University Press, 1991); Helen Milner, *Resisting Protectionism: Global Industries and the Politics of International Trade* (Princeton, NJ: Princeton University Press, 1988); Ronald Rogowski, *Commerce and Coalitions: How Trade Affects Domestic Political Alignments* (Princeton, NJ: Princeton University Press, 1989). See also Peter Gourevitch, *Politics in Hard Times: Comparative Responses to International Economic Crises* (Ithaca, NY: Cornell University Press, 1986).
5. Geoffrey Garrett, *Partisan Politics in the Global Economy* (Cambridge: Cambridge University Press, 1998); Peter A. Hall, *Governing the Economy: The Politics of State Intervention in Britain and France* (New York: Oxford University Press, 1986); Peter J. Katzenstein, ed., *Between Power and Plenty: Foreign Economic Policies of Advanced Industrial Countries* (Madison: University of Wisconsin Press, 1978).
6. Cf. Rogowski, *Commerce and Coalitions*; and Frieden, *Debt, Development, and Democracy*. For a discussion of the utility of comparative analysis at the sectoral (versus national) level, see Herbert Kitschelt, "Industrial Governance Structures, Innovation Strategies, and the Case of Japan: Sectoral or Cross-National Comparative Analysis?" *International Organization* 45, no. 4 (Autumn 1991): 453–93.
7. For more on the importance of linking state/institutional and societal/interest-based variables in the study of trade policy, see Edward Mansfield and Mark

Busch, "The Political Economy of Non-Tariff Barriers: A Cross-National Analysis," *International Organization* 49, no. 4 (Autumn 1995): 723–49; and David Epstein and Sharon O'Halloran, "The Partisan Paradox and the U.S. Tariff, 1877–1934," *International Organization* 50, no. 2 (Spring 1996): 301–24.

8. Good English-language coverage of the emergence of national cinemas in Egypt, Mexico, and India include Jacob Landau, *Studies in the Arab Theater and Cinema* (Philadelphia: University of Pennsylvania Press, 1958); Carl J. Mora, *Mexican Cinema: Reflections of a Society, 1896–2004*, 3rd ed. (Jefferson, NC: McFarland, 2005); Joanne Hershfield and David R. Maciel, eds., *Mexican Cinema: A Century of Film and Filmmakers* (Wilmington, DE: Scholarly Resources, 1999); and Erik Barnouw and S. Krishnaswamy, *Indian Film* (New York: Oxford University Press, 1980).
9. Mancur Olson, *The Logic of Collective Action: Public Goods and the Theory of Groups* (Cambridge, MA: Harvard University Press, 1965).
10. Production, distribution, and exhibition are roughly comparable to manufacturing, marketing, and retail sales in other sectors.
11. Dennis W. Carlton and Jeffrey M. Perloff, *Modern Industrial Organization*, 2nd ed. (New York: HarperCollins, 1994).
12. I derive this corollary in part from the discussion of its obverse: the effects of trade policy on market structure, in Elhanan Helpman and Paul R. Krugman, *Trade Policy and Market Structure* (Cambridge: Massachusetts Institute of Technology Press, 1989), 27–47.
13. Market structure is discussed in Carlton and Perloff, *Modern Industrial Organization*, as well as Helpman and Krugman, *Trade Policy and Market Structure*; and Jean Tirole, *The Theory of Industrial Organization* (Cambridge: Massachusetts Institute of Technology Press, 1988). Olson's work on the collective action problem is evident here. The nature of the film product itself makes for imperfectly competitive markets. See the discussion in Chapter 3, as well as Michael Conant, "The Paramount Decrees Reconsidered," in *The American Film Industry*, rev. ed., ed. Tino Balio, 537–73 (Madison: University of Wisconsin Press), 573.
14. A broad perspective on market institutions is found in Karl Polanyi, "The Economy as Instituted Process," in *Trade and Markets in the Early Empires: Economies in History and Theory*, eds. Karl Polanyi, Conrad M. Arensberg, and Harry W. Pearson, 239–70 (1957; Chicago: Regnery, 1971). See also Polanyi's *The Great Transformation* (1944; Boston: Beacon Press, 1957).
15. The effect of international economic ties on trade preferences was established by Milner, *Resisting Protectionism*.
16. Distributors may be selectively protectionist if their foreign markets differ from the source of import pressure.
17. While mostly addressing the U.S. industry, indirect discussions of industrial organization in the film sector are found in Harold L. Vogel, *Entertainment Industry Economics: A Guide for Financial Analysis*, 6th ed. (Cambridge: Cambridge University Press, 2004); Charles Moul, ed., *A Concise Handbook of Movie Industry Economics* (Cambridge: Cambridge University Press, 2005);

David Throsby, *Economics and Culture* (Cambridge: Cambridge University Press, 2001); Richard E. Caves, *Creative Industries: Contracts Between Art and Commerce* (Cambridge, MA: Harvard University Press, 2002); Hoskins, McFadyen, and Finn, *Global Television and Film*; Kindem, *The American Movie Industry*; and Litman, *The Motion Picture Mega-Industry*. Empirical studies with data on industrial organization elsewhere include Muhammed el-Ashari, *Iqtisadiyat sina'at al-sinima fi Misr: dirasa muqarana* ["Economics of the Film Industry in Egypt: A Comparative Study"] (PhD diss., Cairo University, 1968; Cairo: Dar el-Hana, 1969); Federico Heuer, *La industria cinematográfica mexicana* (Mexico City: privately printed, 1964); and M. A. Oommen and K. V. Joseph, *Economics of Indian Cinema* (New Delhi: Oxford and IBH, 1991).

18. On the Paramount decision by the U.S. Supreme Court, see Ernest Borneman, "United States versus Hollywood: The Case Study of an Antitrust Suit," in Tino Balio, ed., *The American Film Industry*; and Conant, "The Paramount Decrees Reconsidered," in Tino Balio, ed., *The American Film Industry*, 537–74; as well as Michael Conant, *Antitrust in the Motion Picture Industry* (Berkeley: University of California Press, 1960).
19. The voluminous "new institutionalist" literature has expanded rapidly in many directions. For a discussion of the diversity of institutional arrangements in capitalist economies, see J. Rogers Hollingsworth and Robert Boyer, eds., *Contemporary Capitalism: The Embeddedness of Institutions* (Cambridge: Cambridge University Press, 1997).
20. Mostly dealing with "high" culture, discussions of cultural policy in the United States include Dick Netzer, *The Subsidized Muse: Public Support for the Arts in the United States* (Cambridge: Cambridge University Press, 1978); Milton C. Cummings, Jr. and Richard S. Katz, eds., *The Patron State: Government and the Arts in Europe, North America, and Japan* (New York: Oxford University Press, 1987); and Michael Macdonald Mooney, *The Ministry of Culture: Connections Among Art, Money and Politics* (New York: Simon and Schuster, 1980). The distinction between "high" and "popular" culture is a variable social convention.
21. While his focus is on international institutions, John Ruggie draws on Karl Polanyi to argue that international regimes are endowed with both power and "legitimate social purpose," a notion that also may apply to the underlying purpose of state institutions. See John Ruggie, "International Regimes, Transactions, and Change: Embedded Liberalism in the Postwar Economic Order," *International Organization* 36, no. 2 (Spring 1982): 379–415.
22. This conception of state institutions is relational and perhaps an example of what Steven Lukes has called the "third dimension" of power—control over the agenda, or who decides what. While the foregoing is not derived from them, partly useful analyses of cultural policy in Europe include John W. O'Hagan, *The State and the Arts: An Analysis of Key Economic Policy Issues in Europe and the United States* (Cheltenham, UK: Edward Elgar, 1998); Ruth-Blandina M. Quinn, *Public Policy and the Arts: A Comparative Study of Great Britain and Ireland* (Aldershot, UK: Ashgate, 1998); Marla Susan Stone, *The Patron State: Culture and Politics in Fascist Italy* (Princeton, NJ: Princeton University Press,

1998); and David Wachtel, *Cultural Policy and Socialist France* (New York: Greenwood, 1987). A valuable, if dated, series of descriptive country studies of cultural policy worldwide was done by UNESCO. See Magdi Wahba, *Cultural Policy in Egypt* (Paris: UNESCO, 1972).

23. Though not part of my argument, important work on cultural production has been done by the sociologist Pierre Bourdieu. See *The Field of Cultural Production: Essays on Art and Literature* (Cambridge: Polity, 1993). See also Randal Johnson, *The Film Industry in Brazil: Culture and the State* (Pittsburgh, PA: University of Pittsburgh Press, 1987); and Sharon Zukin, *Landscapes of Power: From Detroit to Disney World* (Berkeley: University of California Press, 1991). On the role of the arts in market economies, see Tyler Cowen, *In Praise of Commercial Culture* (Cambridge, MA: Harvard University Press, 1998). A well-known Marxist perspective on the culture industry is that of Theodor Adorno and the Frankfurt School, found in Adorno, *The Culture Industry: Selected Essays on Mass Culture*, ed. J. M. Bernstein (London: Routledge, 1991).
24. Surely, other areas of economic and social policy—such as a state's financial and educational policies—are related to each other: running a deficit to finance school construction has implications for both the state budget and its educational goals. It would be unhelpful, however, to explain educational policy solely in terms of financial constraints or financial policy in terms of educational needs. My argument's implication in this area is that educational policy will be affected by whether a powerful and independent Department of Education exists in a given state structure.
25. Similarly, Helen Milner defines protectionism as "any policy that increases the price of a country's imports or decreases that of its exports." See Milner, *Resisting Protectionism*, 40.
26. I do not employ a quantitative measurement of tariff and nontariff barriers, but relevant definitions of terms and a discussion of the varied forms of trade barriers are found in Sam Laird and Alexander Yeats, *Quantitative Methods for Trade-Barrier Analysis* (New York: New York University Press, 1990).
27. For a sophisticated analysis of French cultural policy in the context of the European Union, see Annabelle Littoz-Monnet, "European Cultural Policy: A French Creation?" *French Politics* 1, no. 3 (November 2003): 255–78. In the developing world, see the repeated references to national identity in the UNESCO country studies of cultural policy cited earlier, as well as official statements by, for example, Nigeria: "The policy shall serve to mobilise and motivate the people by disseminating and propagating ideas which promote national pride, solidarity, and consciousness"; or Pakistan, where official objectives include "To channelize the thoughts and aspirations of our artists, intellectuals, musicians, singers, poets, writers, artisans, architects, stage and film artists, dancers, and other related cultural activities towards the process of national integration." Federal Republic of Nigeria, *Cultural Policy for Nigeria* (Lagos: Federal Government Printer, 1988), 6; and Government of Pakistan, National Commission on History and Culture, *The Cultural Policy of Pakistan* (Islamabad: Pagemaker, 1995), 42.

Chapter 3

1. Hoskins, McFadyen, and Finn characterize nonrivalness as the "joint consumption characteristic" of filmmaking. Colin Hoskins, Stuart McFadyen, and Adam Finn, *Global Television and Film: An Introduction to the Economics of the Business* (Oxford: Clarendon Press, 1997). Observers have long noted nonrivalness in the industry; see, for example, William Victor Strauss, "Foreign Distribution of American Motion Pictures," *Harvard Business Review* 8, no. 3 (April 1930): 307. The other attribute of public goods—"nonexcludability"—is more problematic, since moviegoers must purchase tickets. That said, state policy makers in some contexts do treat film as a kind of public good.
2. Some of these issues are discussed in Albert Moran, "Terms for a Reader: Film, Hollywood, National Cinema, Cultural Identity and Film Policy," in *Film Policy: International, National, and Regional Perspectives*, ed. Moran, 1–19 (New York: Routledge, 1996), 4.
3. Although Ronald Rogowski's discussion of declining transportation costs is part of a broader argument about the political consequences of changes in trade, it is also relevant here. See *Commerce and Coalitions: How Trade Affects Domestic Political Alignments* (Princeton, NJ: Princeton University Press, 1989), 88. My account of distribution costs downplays advertising, which can be a large part of a budget, depending on the release strategy taken by a distributor.
4. Such technological developments include the rise of nontheatrical means of distribution, such as television, video, cable, satellite transmission, DVD, and the Internet. The digital revolution is likely to have substantial effects in this area.
5. For a broad, if somewhat unsystematic, discussion of this, see David Prindle, *Risky Business: The Political Economy of Hollywood* (Boulder, CO: Westview, 1993). The difference between risk and uncertainty is that risk is relatively calculable but uncertainty is not. The prospects for any given film are uncertain, although the industry operates on the basis of calculating and reducing risk.
6. Michael Chanan, *Labour Power in the British Film Industry* (London: British Film Institute, 1976).
7. Harold Vogel claims that six or seven of every ten films "may be broadly characterized as unprofitable." Some of this may be an accounting sleight of hand for tax purposes, but industry insiders echo incessantly the difficulty of achieving financial success for most film projects. Vogel, *Entertainment Industry Economics: A Guide for Financial Analysis*, 2nd ed. (Cambridge: Cambridge University Press, 1998), 31.
8. On the star system, see Cathy Klaprat, "The Star as Market Strategy: Bette Davis in Another Light," in Tino Balio, ed., *The American Film Industry*, rev. ed. (Madison: University of Wisconsin Press, 1985); and Gorham Kindem, "Hollywood's Movie Star System: A Historical Overview," in *The American Movie Industry: The Business of Motion Pictures*, ed. Kindem, 136–45 (Carbondale: Southern Illinois University Press, 1982).
9. The requisite market factors for price discrimination include the capacity to separate markets, some degree of price-setting market power by producers, and

an elasticity of demand that varies across markets. See Hoskins, McFadyen, and Finn, *Global Television and Film*, 69, 72.

10. Hoskins, McFadyen, and Finn argue correctly that the American film industry does not engage in dumping, since it does not sell at a short-term loss to secure a better long-term position in the market. See ibid., 79–80.
11. On the foreign activities of modern, large-scale U.S. firms, see Mira Wilkins, *The Maturing of Multinational Enterprise: American Business Abroad from 1914 to 1970* (Cambridge, MA: Harvard University Press, 1974). The early history of U.S. government support to the film sector is covered judiciously in John T. Trumpbour, *Selling Hollywood to the World: U.S. and European Struggles for Mastery of the Global Film Industry, 1920–1950* (Cambridge: Cambridge University Press, 2002). A prominent argument that power itself is "becoming less fungible, less coercive, and less tangible," is Joseph S. Nye Jr., *Bound to Lead: The Changing Nature of American Power* (New York: Basic Books, 1990), 188–201 passim.
12. A general work tracing this long-standing conflict through the last round of negotiations for the General Agreement on Tariffs and Trade is David Puttnam, *The Undeclared War: The Struggle for Control of the World's Film Industry* (New York: HarperCollins, 1997). Puttnam is an experienced British producer (*Chariots of Fire, The Killing Fields*) who ran Columbia Pictures for a short period.
13. The most sophisticated consideration of language markets is found in Hoskins, McFadyen, and Finn, *Global Television and Film*, 27–36. A more partisan economic analysis of "linguistic markets" is Steven S. Wildman and Stephen E. Siwek, *International Trade in Films and Television Programs* (Cambridge, MA: Ballinger, 1988).
14. American exports are not pure profit, with production costs amortized domestically. As the historical record shows, the U.S. industry has long relied on foreign markets to defray production costs. See the letter dated February 18, 1957, from Motion Picture Export Association vice president G. Griffith Johnson to J. M. Colton Hand, Division of Commercial Policy, U.S. Department of State, Confidential U.S. State Department Central Files 1955–1959, Record Group 59, Reel 7, Decimal Number 611.74231/2-1857. Johnson notes that since "nearly half of the gross income earned by our companies from film distribution is received from markets outside of the United States . . . a comparable proportion of actual film production costs must be allocated to these markets."
15. Powerful non-American firms like Pathé and Gaumont (France), UFA (Germany), Rank (Britain), and Toho (Japan) were influential in certain contexts. For a comparative and historical look at competition in other globally oriented industries, see Alfred D. Chandler Jr., "The Evolution of Modern Global Competition," in *Competition in Global Industries*, ed. Michael E. Porter, 405–48 (Boston: Harvard Business School Press, 1986).
16. On the early development and technological aspects of filmmaking, see A. R. Fulton, "The Machine" in Balio, ed., *The American Film Industry*, 27–42; and Raymond Fielding, ed., *A Technological History of Motion Pictures and*

Television: An Anthology from the Pages of the Journal of the Society of Motion Picture and Television Engineers (Berkeley: University of California Press, 1967).

17. Good coverage of this early period, focusing on U.S. distribution power in non-European markets, is Kristin Thompson, *Exporting Entertainment: America in the World Film Market, 1907–34* (London: British Film Institute, 1985).
18. Thompson, *Exporting Entertainment*, 2; and Balio, "Part II: Struggles for Control," in Balio, ed., *The American Film Industry*, 122.
19. For more on the Prince of Wales's speech and an early American perspective on the matter, see Edward G. Lowry, "Trade Follows the Film," *Saturday Evening Post* 198 (November 7, 1925): 12. The latter phrase was invoked repeatedly in subsequent decades in discussions of the purportedly extraordinary effects of American motion pictures on demand for American goods and services.
20. A sympathetic but useful early account of the MPPDA is Raymond Moley, *The Hays Office* (New York: Bobbs-Merrill, 1945), esp. chap. 15, "Foreign Relations," 169–86.
21. On the role of corporatism, beginning especially with the Hoover administration, see Trumpbour, *Selling Hollywood to the World*, 17–63.
22. Before being reconstituted as the Motion Picture Export Association in 1945, the Foreign Department was renamed the International Department in 1943 because "we wanted to emphasize the fact that we considered that motion pictures had become a vital, almost universal, international medium of communication, and that no nation was 'foreign' to their sphere of influence." Will H. Hays, *The Memoirs of Will H. Hays* (Garden City, NY: Doubleday, 1955), 505; Thomas Guback, "Hollywood's International Market," in Balio, ed., *The American Film Industry*, 463–86.
23. On the nature and development of this style, as well as its relationship to the economic system in which films were made, see David Bordwell, Janet Staiger, and Kristin Thompson, *The Classical Hollywood Cinema: Film Style and Mode of Production to 1960* (New York: Columbia University Press, 1985); and Ruth Vasey, *The World According to Hollywood, 1918–1939* (Madison: University of Wisconsin Press, 1997).
24. While he develops a "media imperialism" thesis, Jeremy Tunstall addresses a similar process of American media dominance in filmmaking, newspapers, and commercial publishing; see Tunstall, *The Media are American: Anglo-American Media in the World* (New York: Columbia University Press, 1977).
25. On the Mexican industry's development in reflection of Hollywood, see Charles Ramírez Berg, *Cinema of Solitude: A Critical Study of Mexican Film, 1967–1983* (Austin: University of Texas Press, 1992), 16–17. Egyptian observers repeatedly have referred to their industry with phrases like "the Hollywood of the East." Other examples of this emulative tendency include India's Bombay cinema—"Bollywood"—and even the name given to a film studio built in 1936 England by J. Arthur Rank: Pinewood Studios.
26. There are specific path dependencies in the complex technological development of the industry, and elements of filmmaking that are common today were neither necessary nor inevitable. The width of film stock and the spacing between

sprocket holes, for example, became a universally accepted standard that aided in the global spread of American motion pictures, even if they originated in random choices. See Robert Sklar, *Movie-Made America: A Cultural History of American Movies* (New York: Random House, 1975), 216.

27. The five leading majors firms that dominated the industry by 1930 were Paramount, Warner Brothers, 20th Century-Fox, Metro-Goldwyn-Mayer (MGM), and Radio-Keith-Orpheum (RKO); three associated "minors" included Columbia, United Artists, and Universal Pictures.
28. A good guide to the workings of the major studios in this period is Douglas Gomery, *Hollywood Studio System* (New York: St. Martin's, 1986). Anticompetitive practices are discussed extensively in Mae D. Huettig, *Economic Control of the Motion Picture Industry: A Study in Industrial Organization* (Philadelphia: University of Pennsylvania Press, 1944).
29. On the introduction of sound, see Douglas Gomery, "The Coming of Sound to the American Cinema: A History of the Transformation of an Industry" (PhD diss., University of Wisconsin at Madison, 1975).
30. The challenges facing European filmmakers, and the advantages held by the United States, were apparent very early. See Strauss's article from 1930, "Foreign Distribution of American Motion Pictures," *Harvard Business Review*, 307–15. A first-rate account of British-American rivalry in the film trade is Ian Jarvie, *Hollywood's Overseas Campaign: The North Atlantic Movie Trade, 1920–1950* (Cambridge: Cambridge University Press, 1992).
31. Trumpbour, *Selling Hollywood to the World*, 183–99.
32. Richard A. May, American Trade Commissioner, to J. Morton Howell, American Minister, March 13, 1926. National Archives, *Records of the Department of State Relating to the Internal Affairs of Egypt, 1910–29*, 883.40 Social Matters. The fact that films were shipped from the United States to Europe, and on to the Middle East and elsewhere via Egypt, indicates that trade statistics need to be handled with caution, since they often refer only to the country of immediate origin. As May noted in the source above, "The official import returns do not show that the United States enjoys [a] large proportion of the business for the reason that many American films are imported into Egypt from France, Italy, and other countries after having been showed there."
33. *Cine Film*, special edition, no. 26 (June 1950): 26. The French company, Gaumont, introduced the idea of a percentage deal, instead of pricing film rentals by their length, and American companies followed suit.
34. C. J. North, "Our Foreign Trade in Motion Pictures," in "The Motion Picture in its Social and Economic Aspects," *Annals of the American Academy of Political and Social Science* 128 (November 1926): 107. North was head of the U.S. Commerce Department's Motion Picture Section.
35. Figure 3.1 underrepresents American films for reasons noted earlier. It has been calculated using the nominal value of film imports, as listed in the *Annual Statement of the Foreign Trade of Egypt*, Ministry of Finance, Statistical Department, 1918–1960.

36. Carl E. Milliken, Manager, International Department, MPAA to George Canty, Assistant Chief, Telecommunications Division, Department of State. 883.4061 Motion Pictures/3-1346 and enclosed "Confirmation of Cable," RKO Radio Pictures Inc., Export Division.
37. The latter film was banned by Egyptian censors in 1938 for depicting the French revolution. John Eugene Harley, *World-Wide Influences of the Cinema: A Study of Official Censorship and the International Cultural Aspects of Motion Pictures* (Los Angeles: University of Southern California Press, 1940), 121.
38. See Magda Wassef, ed., *Egypte: 100 ans de cinéma* (Paris: Editions Plume and Institut du Monde Arabe, 1995), 20–21.
39. When the Egyptian industry itself celebrated its twenty-first anniversary in 1948, it dated its birth to the 1927 production of *Leila*. See *al-Film Cine-Orient*, no. 8 (October 16, 1948). Official Egyptian historiography usually downplays the role of foreign nationals in the development of the industry, especially after 1927. The early history of Egyptian filmmaking is found in Ahmed al-Hadari, *Tarikh al-sinima fi Misr: al-juz al-awwal min bidayat 1896 ila akher 1930* [*A History of the Cinema in Egypt: Part One, from the Beginning of 1896 to the End of 1930*] (Cairo: Nadi al-Sinima bil-Qahira, 1989).
40. Nathan D. Golden, *Review of Foreign Film Markets during 1936*, U.S. Department of Commerce, Bureau of Foreign and Domestic Commerce, Motion Picture Section (Washington, DC: April 1937), 64. In 1934, imported sound films came to Egypt in large numbers. This is evident from the rapidly changing balance between sound and silent film imports in the early 1930s:

Year	Silent	Sound
1931	97%	3%
1932	93%	7%
1933	77%	23%
1934	2%	98%
1935	1%	99%

Source: Calculated from Annual Statement of the Foreign Trade of Egypt, various years.

41. May to Howell, March 13, 1926. For data on extensive Egyptian re-exports, see the transit trade tables in the *Annual Statement of the Foreign Trade of Egypt*. This source also records the substantial movement of film between Egypt and the Sudan, almost certainly destined for British troops stationed there.
42. On Tal'at Harb and the founding of Studio Misr, see Ilhami Hasan, *Muhammad Tal'at Harb: Ra'id sina'at al-sinima al-misriyya, 1867–1941* [*Muhammad Tal'at Harb: Leader of the Egyptian Film Industry, 1867–1941*] (Cairo: General Egyptian Book Organization, 1986).
43. The sources for Figure 3.3 are the following: Federation of Egyptian Industries, *Year Book 1973* (Cairo: General Organization for Government Printing Offices, 1973), 231; Ministry of Culture, Technical Office for the Cinema, *Sina'at al-sinima: haqa'iq w-arqam* [*The Cinema Industry: Facts and Figures*] (Cairo: General Egyptian

Organization for the Cinema, Radio, and Television, various years); Ali Abu Shadi, "Chronologie: 1896–1994," in Wassef, ed., *Egypte: 100 ans de cinéma*, 18–39. For a list of more than 1,500 feature films made in Egypt from 1927 to 1973, see 'Abd al-Min'em Sa'ad, *al-Sinima al-misriyya fi mawsim 1973*, 218–46.

44. Studio details are in Jacques Pascal, ed., *The Middle East Motion Picture Almanac, 1946–47*, 1st ed. (Cairo: S.O.P. Press, 1947), 127–32. Names and details on the explosion of production companies are listed in Galal al-Sharqawi, *Risala fi tarikh al-sinima al-'arabiyya* [*A Treatise on the History of the Arab Cinema*] (Cairo: al-Misriyya, 1970), 101–5; translation of PhD diss., Institut des hautes études cinématographiques, 1962.
45. The popularity of the cinema is noted by M. M. Mosharrafa, *Cultural Survey of Modern Egypt*, Part II (London: Longmans, Green, 1948), 59.
46. Jacob M. Landau, *Studies in the Arab Theater and Cinema* (Philadelphia: University of Pennsylvania Press, 1958), 179–80.
47. Studio rental cost between £E 1,500 and £E 4,000 per month in 1947. Cost breakdowns are in Pascal, *The Middle East Motion Picture Almanac*, 111. On foreign distribution, see Karen Finlon Dajani, "Egypt's Role as a Major Media Producer, Supplier and Distributor to the Arab World: An Historical-Descriptive Study" (PhD diss., Temple University, 1979), esp. 130–42.
48. For an overview of Egypt's star system, see Christophe Ayad, "Le star-système: De la splendeur au voile," in Wassef, ed., *Egypte: 100 ans de cinema*, 134–41. An indication of the centrality of Egyptian stars to the industry is their presentation, along with a handful of directors, in Mustapha Darwish, *Dream Makers on the Nile: A Portrait of Egyptian Cinema* (Cairo: American University in Cairo, 1998). The limited number of Egyptian stars is discussed and quantified by el-Sharqawi, *Risala fi tarikh al-sinima al-'arabiyya*, 105–8. On genres, see Ali Abu Shadi, "Genres in Egyptian Cinema," in *Screens of Life: Critical Film Writing from the Arab World*, vol. 1, ed. and trans. Alia Arasoughly, 84–129 (Quebec: World Heritage Press, 1996).
49. Paulo Antonio Paranaguá, *Mexican Cinema* (London: British Film Institute and IMCINE, 1995), 24–25.
50. Hays, *The Memoirs of Will H. Hays*, 333–34.
51. North, "Our Foreign Trade in Motion Pictures," 101.
52. For a detailed study of the involvement of United Artists in Latin America from 1919 to 1951, based on corporate records, see Gaizka S. de Usabel, *The High Noon of American Films in Latin America* (Ann Arbor, MI: UMI Research Press, 1982).
53. Carl J. Mora, *Mexican Cinema: Reflections of a Society, 1896–1988*, rev. ed. (Berkeley: University of California Press, 1989), 31–32.
54. John King, *Magical Reels: A History of Cinema in Latin America* (New York: Verso, 1990), 32; Dudley Andrew, *Mists of Regret: Culture and Sensibility in Classic French Film* (Princeton, NJ: Princeton University Press, 1995), 96–98.
55. Mora, *Mexican Cinema*, 32.
56. For more on Hollywood's influence and the "apprenticeship of dreams" by one of Mexico's leading cultural critics, see Carlos Monsiváis, "Mexican Cinema: Of

Myths and Demystifications," in *Mediating Two Worlds: Cinematic Encounters in the Americas*, ed. John King, Ana M. López, and Manuel Alvarado, 139–46 (London: British Film Institute, 1993), 141–42.

57. Golden, *Review of Foreign Film Markets during 1936*, 109. For more on the war's early effects on the film trade worldwide see the chapter, "International Commerce in Films," in Harley, *World-Wide Influences of the Cinema*, 245–67.
58. See Thomas Doherty, *Projections of War: Hollywood, American Culture, and World War II* (New York: Columbia University Press, 1993), 299–304.
59. Emilio García Riera, *Historia documental del cine mexicano, Vol. 1: 1929–1937* (Guadalajara and Mexico City: University of Guadalajara and National Council for Culture and the Arts, 1992), 211.
60. Emily S. Rosenberg, *Spreading the American Dream: American Economic and Cultural Expansion, 1890–1945* (New York: Farrar, Straus, and Giroux, 1982), 206–8.
61. Jorge Schnitman, *Film Industries in Latin America: Dependency and Development* (Norwood, NJ: Ablex, 1984).
62. Emilio García Riera, *Historia documental del cine mexicano, Vol. 5: 1949–1950* (Guadalajara: University of Guadalajara; Mexico City: National Council for Culture and the Arts, 1992), 7.
63. See Michael Conant, *Antitrust in the Motion Picture Industry: Economic and Legal Analysis* (Berkeley: University of California Press, 1960); and Conant, "The Paramount Decrees Reconsidered," in Balio, ed., *The American Film Industry*, 537–73.
64. The majors were not forced to break ties specifically with the exhibition subsector; they were given a choice as to which parts of the business to retain. Like John D. Rockefeller and Standard Oil decades earlier, they recognized the immense power of distribution and held on to it.
65. Balio, *The American Film Industry*, 401.
66. William F. Hellmuth Jr., "The Motion Picture Industry," in *The Structure of American Industry: Some Case Studies*, rev. ed., ed. Walter Adams (New York: Macmillan, 1954), 380.
67. William F. Hellmuth Jr., "The Motion Picture Industry," in *The Structure of American Industry: Some Case Studies*, 3rd ed., ed. Walter A. Adams (New York: Macmillan, 1961), 410. Hellmuth notes that U.S. foreign earnings in the film sector were $300 million in 1958, of which approximately $210 million were remittable to the United States.
68. An important study of these trends throughout the 1950s and 1960s is Guback, *The International Film Industry*.
69. On American firms gaining access to European subsidies, see Guback, *The International Film Industry*, 166–70.
70. Another piece of equipment for location shooting was the "Cinemobile," a mobile studio invented by Egyptian-born cinematographer, Fouad Said, and still used today. This and other technological changes are covered by Robert H. Stanley, *The Celluloid Empire: A History of the American Movie Industry* (New York: Hastings, 1978), 242 and passim.

71. Only DeMille's film was a commercial success. After seeing the Howard Hawks picture, the Ministry of National Guidance reportedly instructed that it no longer be exhibited in Egypt. *Cine Film*, no. 91 (December 1955): 10. Interesting contemporaneous Egyptian accounts are in "Hollywood on the Nile," *The Egyptian Economic and Political Review* 1, no. 1 (September 1954): 24–25; and "The Ten Commandments," *The Egyptian Economic and Political Review* 1, no. 3 (November 1954): 12. DeMille's autobiography has additional details, as does *Cine Film*, no. 78 (November 1954): 7, which mentions the huge set constructed for the picture at Beni Youssef and tells how DeMille was received by President Nasser and feted at the Zamalek Officers' Club by one of the Free Officers, Waguih Abaza, head of the Nile Cinema Company.
72. This increase may have been undermined by the declining demand for films in some Middle Eastern countries due to the departure of tens of thousands of European residents and military personnel. Georges Sadoul, "Geography of the Cinema and the Arab World," in *The Cinema in the Arab Countries*, ed. Georges Sadoul, 129–36 (Beirut: UNESCO and Interarab Centre of Cinema and Television, 1966), 132–33.
73. This figure is based on data from the Central Agency for Public Mobilization and Statistics (CAPMAS), *al-Ihsa'at al-thaqafiyya: al-sinima wal-masrah* [*Cultural Statistics: The Cinema and Theater*], 1964–1996; and Ministry of Culture, *Sina'at al-sinima: haqa'iq w-arqam* [*The Cinema Industry: Facts and Figures*], various years.
74. *Cine Film*, no. 33 (February 1951): 15. For their part, Israeli government censors tried to avoid seeming partisan by banning both Soviet propaganda movies and U.S. anticommunist pictures like *The Iron Curtain, Conspirators*, and *The Red Danube*. American companies are said to have complained, asking the Israelis to allow the public to make their own judgments in the ideological contest. *Cine Film*, no. 30 (November 1950): 11.
75. See the correspondence in mid-1951 between Herbert T. Edwards, Chief, International Motion Picture Division, Department of State and Theodore Smith of the MPAA regarding the proposed dubbing of an Arabic-language version of the anticommunist picture, *Woman on Pier 13*. U.S. officials feared that its propaganda value might backfire when presented to an Egyptian audience, since the hero was a communist murderer and American society was portrayed as helpless. [874.452/5-2151] On the remittance issue, see the correspondence in early 1957 between the MPAA's Griffith Johnson and officials in the State and Treasury Departments, who demonstrated considerable autonomy in opposing the MPAA's efforts to access Egyptian funds frozen in the United States after the Suez crisis. [611.74231/3-157] This is clear in the correspondence in late 1952 between J. M. Colton Hand of State's Office of Commercial Policy and John G. McCarthy, the MPAA vice president and head of the International Division, who wrote to alert State to the possible formation of an Egyptian national film center. The latter development was seen as less threatening by U.S. officials than by their counterparts in the film business, with State's Colton Hand pointing out in his response that "many [Egyptian] producers are beset by financial

difficulties" and therefore naturally were seeking government assistance. [874.452/11-2852]

76. Hellmuth, "The Motion Picture Industry," 381–82. He further notes here the great lengths to which American companies went to remit their earnings by, for example, "increasing production abroad and by acquiring foreign theater holdings . . . [and buying] gold nuggets, wines, marble, and other commodities abroad [to] obtain dollars by selling these in the United States."
77. For more on Egypt's postwar balance of payments and foreign trade position, see Charles Issawi, *Egypt at Mid-Century: An Economic Survey* (New York: Oxford University Press, 1954); and Bent Hansen and Karim Nashashibi, *Foreign Trade Regimes and Economic Development: Egypt* (New York: National Bureau of Economic Research and Columbia University Press, 1975).
78. See the correspondence between Gerald M. Mayer, Managing Director, International Division, MPAA and George Canty, Assistant Chief, Division of Commercial Policy, U.S. State Department, 883.40061 MP/5-547 and 5-747.
79. Cairo Airgram A-719 dated June 28, 1949, from H. G. Minigerode/American Embassy to SecState, 883.4061 M.P./6-2249.
80. For more on the cultural dimension of the cold war in areas like education policy, see Frank A. Ninkovich, *The Diplomacy of Ideas: U.S. Foreign Policy and Cultural Relations, 1938–1950* (New York: Cambridge University Press, 1981). The origins of U.S. policy are discussed by Rosenberg, *Spreading the American Dream*. The first evidence of cold war tensions being played out in the film industry in Egypt came in November 1948, when the American Consul General in Alexandria reported that the manager of a theater showing the anticommunist picture, *The Iron Curtain*, terminated the screening after receiving a letter threatening to blow up the theater if it continued. See Buell to State, November 9, 1948, 883.4061 MP/11-948. Egypt's anticommunist monarchy did not permit Soviet or Eastern bloc film imports; a year after the July revolution, the first Soviet films reportedly were to be imported and distributed for showing in August 1953. See McKee to State, July 28, 1953, 874.452/7-2853.
81. On the Soviet cartoon, see Sklar, *Movie-Made America*, 215. The seminal Marxist take on Donald Duck is Ariel Dorfman and Armand Mattelart, *How to Read Donald Duck: Imperialist Ideology in the Disney Comic*, trans. and rev. ed. (New York: International General, 1991).
82. Such claims became more pronounced when U.S. exports came under greater economic pressure from the Egyptians. Representatives may have been successful, judging by the number of "info copies" of cables forwarded to the CIA (17, in one instance, indicating substantial and bureaucratically diverse interest in this issue). See Haring to State, April 11, 1957, 874.452/4-1157.
83. See the confidential cable from W.H. Weathersby to State, October 20, 1956, 874.452/10-2056 on the subject of "Communist Penetration of Egyptian Theatrical Film Market," and the joint State-USIA cable, March 18, 1957, 874.452/3-1857 on Soviet attempts to buy or lease theaters in Cairo, Alexandria, and Port Said.

84. Tilmissani directed one of Egypt's classic works of cinematic realism: *The Black Market* (1945). A U.S. Embassy cable on "Soviet Bloc Activities in the Motion Picture Trade," dated May 10, 1957 [874.452/5-1057], refers to what probably is this book, though the title is said to be "American Ambassador in Technicolor and Cinemascope" and authorship is attributed to the Soviets. Robert Vitalis first unearthed and discussed this and the related diplomatic correspondence in a December 1995 conference paper, later published as "American Ambassador in Technicolor and Cinemascope: Hollywood and Revolution on the Nile," in *Mass Mediations: New Approaches to Popular Culture in the Middle East and Beyond*, ed. Walter Armbrust, 269–91 (Berkeley: University of California Press, 2000). Egyptian film critic Samir Farid maintains that the words "Natural Colors" in Tilmissani's book title refer specifically to the increasingly popular color films of the mid-1950s, and that Tilmissani's book could be read in conjunction with its right-wing 1936 counterpart: Ahmad Badrakhan's *al-Sinima* [*The Cinema*] (Cairo: al-Halabi Press, 1936). Ironically, the financial failure of *The Black Market* led Tilmissani to make more commercially oriented films.
85. Charles Issawi notes that "By the end of the war Egyptian individuals and institutions had accumulated sterling balances of about £E 400 million." Issawi, *Egypt at Mid-Century*, 204. In a 1947 edition of this book written in 1942–1943, he estimated annual film industry expenditures at £E 500,000; filmmaking was widely viewed then as a lucrative business, with Egypt having "possibilities as a first-rate international film centre." Issawi, *Egypt: An Economic and Social Analysis* (London: Oxford University Press, 1947), 93. The so-called war profiteers, however, had no interest in filmmaking as anything other than a short-term investment. See Sa'ad el-Din Tawfiq, *Qissat al-sinima fi Misr: dirasa naqdiyya* [*The Story of the Cinema in Egypt: A Critical Study*] (Cairo: Dar el-Hallal, 1969), 83–84.
86. *Annual Statement of the Foreign Trade of Egypt*, various years. Ten Egyptian films were exported to Brazil in 1948. *Cine Film*, no. 26 (June 1950): 63.
87. *Annual Statement of the Foreign Trade of Egypt, 1954*, 570–71. For more on Egypt's early foreign markets, see *Cine Film*, special edition, no. 66 (November 1953): 8–12.
88. *Annual Statement of the Foreign Trade, 1953*, 147. The government record is useful for tracking the relative standing of the industry, particularly from when it first appears in the ranked list of exports in 1943 (at no. 34) through its high-point in 1949 (at no. 21) to its precipitous drop in 1960 (to no. 77).
89. Landau (180) gives the figure of £E 2.7 million, citing *Ruz al-Yousef*, July 26, 1954. Actual profits after taxes and other costs are impossible to determine with accuracy. Jacques Pascal, editor of the trade journal *Cine Film*, regularly referred to filmmaking as Egypt's second industry after cotton; see, for example, *Cine Film*, no. 18 (September 1949): 1–2. It is more likely that the industry enjoyed several spectacularly profitable years but was unable to establish itself permanently on such a lucrative basis.

90. Samir Farid, "Periodization of Egyptian Cinema," in *Screens of Life: Critical Film Writing from the Arab World*, ed. and trans. Alia Arasoughly, 1–18 (Quebec: World Heritage Press, 1996), 11.
91. *Sina'at al-sinima: haqa'iq w-arqam* [*The Film Industry: Facts and Figures*], General Egyptian Organization for the Cinema, Radio, and Television, Technical Office for the Cinema, March 1964, p. 14. Inter-Arab rivalry of the period is covered in Malcolm Kerr, *The Arab Cold War: Gamal 'Abdel Nasser and His Rivals*, 3rd ed. (New York: Oxford University Press, 1971); and Stephen M. Walt, *The Origins of Alliances* (Ithaca, NY: Cornell University Press, 1987).
92. See *Sina'at al-sinima: haqa'iq w-arqam* for details on the decline. Since it had no public cinemas, Saudi imports were for private showing only.
93. *International Motion Picture Almanac, 1953–54*, 818.
94. *International Motion Picture Almanac, 1956*, 794, and *1957*, 850. The peso was devalued from 8.65 to 12.50 to the dollar in 1954. Hollywood's work in Durango is noted by Emilio García Riera, *Historia del cine mexicano* (Mexico City: Secretaría de Educacíon Pública, 1986), 244.
95. American filmmakers also shot on location in Mexico, such as when making Elia Kazan's *Viva Zapata* in 1952.
96. King, *Magical Reels*, 129–30.
97. Brazil's Portuguese-language films had a linguistic disadvantage, though this did not preclude exports, especially in later years with state patronage. See Randal Johnson, *The Film Industry in Brazil: Culture and the State* (Pittsburgh: University of Pennsylvania Press, 1987), 138–42.
98. Emilio García Riera, *Historia del cine mexicano*, 221.
99. Federico Heuer, *La Industria cinematográfica mexicana* (Mexico City: privately printed, 1964), 86–87. Heuer was the director of the Banco Nacional Cinematográfico when he wrote this detailed study.
100. John King, "Mexico: Inside the Industrial Labyrinth," in King, *Magical Reels*, 129–44. See also Fernando Macotela Vargas, *La industria cinematográfica mexicana: estudio juridico y economico* (Mexico City: National Autonomous University of Mexico, 1969). Vargas served as director of the *Cineteca Nacional* in the 1970s.
101. Balio, "Adjusting to the New Global Economy: Hollywood in the 1990s," in Moran, ed., *Film Policy*, 27. The initial merger and acquisition activity included Transamerica's purchase of United Artists in 1967, Kinney National Service's buying of Warner Brother in 1969, and MGM's expansion into hotels and gambling. For more on these and subsequent merger and acquisition activity, see the organigrams in Vogel, *Entertainment Industry Economics*, 57; Allen J. Scott, "A New Map of Hollywood and the World," University of California, Los Angeles, unpublished manuscript; and Lisa Mirabile, ed., *International Directory of Company Histories*, vol. 2 (Chicago: St. James, 1990). While out of date in important ways, comprehensive organigrams depicting decades of ownership change by the majors are in Barry Langford and Douglas Gomery, "Studio Genealogies: A Hollywood Family Tree," *Gannett Center Journal* 3, no.

3 (Summer 1989): 104–22. Useful if dated discussion of some of the larger trends is also in Stanley, *The Celluloid Empire*, 231–41, 252–67.

102. In the 1980s, Hollywood was able to increase its vertical integration under the Reagan-era interpretation of antitrust laws. Some sense of the larger trends in profits and losses can be gained from the table, "Studio Revenues and Profits, 1920–1986," in *The Hollywood Story*, by Joel W. Finler (New York: Crown, 1988), 286–87.

103. Foreign state subsidies eventually came to support even large-scale "American" productions like Steven Spielberg's *Saving Private Ryan*, which obtained Irish government support for filming the invasion sequence on the Irish coast and used several hundred Irish army extras. In terms of the internationalization of ownership, Vogel notes (38–39) that Japan's Sony took control of Columbia and TriStar; Canada's Seagram's gained an 80 percent share in Universal; and Australia's News Corporation acquired Twentieth Century-Fox. More recently, France's Vivendi Universal took over both Canal Plus and Seagram's, which gave it control of Universal Studios. Also, the London-based United International Pictures obtained foreign distribution rights for Universal, Paramount, and MGM/United Artists. These examples notwithstanding, one should not discount the continued importance of American firms in all aspects of the industry, notably the key, related segments of financing and distribution. For an historical perspective on American dominance, see Thomas H. Guback, "Film as International Business: The Role of American Multinationals," in Kindem, *The American Movie Industry*, 336–50.

104. On the changing forms of production up to 1960, including the "package-unit," "producer-unit," and "director-unit" systems, see Bordwell, Staiger, and Thompson, *The Classical Hollywood Cinema*.

105. Michael Storper, "The Transition to Flexible Specialisation in the U.S. Film Industry: External Economies, the Division of Labour, and the Crossing of Industrial Divides," *Cambridge Journal of Economics* 13, no. 3 (September 1989): 273–305; Michael Storper and Susan Christopherson, "Flexible Specialization and Regional Industrial Agglomerations: The Case of the U.S. Motion Picture Industry," *Annals of the Association of American Geographers* 77, no. 1 (1987): 104–17. See also the critique of Storper's work in Asu Aksoy and Kevin Robins, "Hollywood for the 21st Century: Global Competition for Critical Mass in Image Markets," *Cambridge Journal of Economics* 16, no. 1 (March 1992): 1–22; and Storper's response, "Flexible Specialisation in Hollywood: A Response to Aksoy and Robins," *Cambridge Journal of Economics* 17 (1993): 479–84. Useful discussion of flexible specialization in general is found in Paul Hirst and Jonathan Zeitlin, "Flexible Specialization: Theory and Evidence in the Analysis of Industrial Change," in *Contemporary Capitalism: The Embeddedness of Institutions*, ed. J. Rogers Hollingsworth and Robert Boyer, 220–39 (Cambridge: Cambridge University Press, 1997).

106. Storper and Christopherson, "Flexible Specialization and Regional Industrial Agglomerations," 108. An "independent" is a production company unaffiliated with a major studio-distributor.

107. Storper, "The Transition to Flexible Specialisation in the U.S. Film Industry," 298–99. Flexible specialization in the U.S. industry only came into being because of the domestic conditions that encouraged its development.
108. CAPMAS, *al-Ihsa'at al-thaqafiyya*, various years; and Ministry of Culture, *Sina'at al-sinima*, various years.
109. Welles never actually filmed in Egypt, but the fact that the director of *Citizen Kane* came close to doing so says something about Egypt's impressive standing in the industry in the late 1940s. For more on the Welles contract, see "Echos et Nouvelles: Orson Welles va tourner en Egypte," *Cine Film*, no. 19 (November 1949): 7. In more recent years, well-known Hollywood productions ostensibly set in Egypt, such as *The English Patient*, have been filmed in Morocco and Tunisia instead due to the political obstacles and bureaucratic inefficiencies for which the Egyptian industry has become known.
110. Madkour Sabet, "L'économie de l'industrie du cinéma," in Wassef, *Egypte: 100 ans de cinéma*, 131.
111. The rent-seeking efforts of the Cinema Industry Chamber [*Ghurfa sina'at al-sinima*], an affiliate of the Federation of Egyptian Industries, calls into question the reliability of their data, but the Chamber's annual reports discuss many of these matters. See *al-Taqrir al-sanawi* [Annual Report], various years. For the Cinema Chamber's views on piracy in the 1970s, see *Yearbook 1981*, Federation of Egyptian Industries, 541–46.
112. Dajani, *Egypt's Role*, 142.
113. Sadoul, *Cinema in the Arab Countries*, 132.
114. Madkour Sabet, "L'économie de l'industrie du cinéma," 132.
115. *World Communications: A 200-Country Survey of Press, Radio, Television, and Film* (Paris: UNESCO, 1975), 205.
116. Berg, *Cinema of Solitude*, 12.
117. Paranaguá, *Mexican Cinema*, 54.
118. Mora, *Mexican Cinema*, 139. For more on the Echeverría period, see Paola Costa, *La "Apertura" cinematográfica: México, 1970–1976* (Puebla, Mexico: Universidad Autónoma de Puebla, 1983).
119. As interior minister in 1968, Echeverría was widely considered responsible for the massacre. For more on the effects of 1968 on state relations with intellectuals, see Roderic A. Camp, *Intellectuals and the State in Twentieth-Century Mexico* (Austin: University of Texas Press, 1985), 208–12.
120. For details, see the country listings in *World Communications*, 137–251.
121. The estimate of receipts is from Mora, 139, citing Manuel Ampudia Girón, then president of CANACINE, the National Chamber of the Cinematic Industry.
122. Critical successes like *Amores Perros*, *Y Tu Mama Tambien*, *Pan's Labyrinth*, and *Babel* highlight the resilience of the industry and its most talented artists.

Chapter 4

1. Dennis W. Carlton and Jeffrey M. Perloff, *Modern Industrial Organization*, 2nd ed. (New York: HarperCollins, 1994).

2. Four women led early production: Aziza Amir, Asia Dagher, Fatima Rushdi, and Mary Queeny. Key distributors of Egyptian films included the Misr Company for the Theater and Cinema, Benha Films, Nahas Films, Charles Lifschitz, and Cairo Films; leading locally owned distributors handling foreign films included Alma (originally Prosperi) Film and Josy Film, as well as Ideal Motion Pictures and Dollar Film. *Cine Film*, no. 26 (June 1950): 26.
3. Ahmad al-Hadari, *Tarikh al-sinima fi Misr: al-juz al-awwal min bidayat 1896 ila akhir 1930* [*History of the Cinema in Egypt: Part One, from the Beginning of 1896 to the End of 1930*] (Cairo: Nadi al-Sinima bil-Qahira, 1989).
4. *Sina'at al-sinima: haqa'iq w-arqam* [*The Film Industry: Facts and Figures*], General Egyptian Organization for the Cinema, Radio, and Television, Technical Office for the Cinema, March 1964, p. 2.
5. Ministry of Culture: Cinema, Theatre and Music Organization, *The Motion Picture Industry in Egypt*, May 1979, p. 2.
6. A contemporaneous account of the industry's wartime expansion and post-war decline is in *Cine Film*, no. 11 (December 1, 1948): 3. The number of spectators reportedly grew by 245 percent during the war. *Cine Film*, no. 15 (March 1949): 22.
7. A classified ad in an industry trade journal in 1948 provides a telling indication of the ad hoc nature of film financing in this period: "On demande capitaliste disposant de £E 1,000 ou 1,500 pour excellente affaire cinématographique. Immobilisation six mois. Affaire assurée. Ecrire à la direction du journal." *Cine-Orient* 1, no. 3 (July 1948): 6.
8. *The Middle East Motion Picture Almanac of 1946–47* (111) states there were 140 production companies in Egypt. The vast majority could not have been operating actively, since Galal al-Sharkawi's list of films from the period includes only thirty producers affiliated with films released that year. The discrepancy is probably not an error so much as an indication of the sporadic involvement of individual investors in that period, and therefore of a fragmented and competitive market. See Galal al-Sharkawi, *Risala fi tarikh al-sinima al-'arabiyya* (Cairo: al-Misriyya, 1970).
9. Bent Hansen and Karim Nashashibi, *Foreign Trade Regimes and Economic Development: Egypt*, vol. 4, Special Conference Series on Foreign Trade Regimes and Economic Development, National Bureau of Economic Research (New York: Columbia University Press, 1975), 3–5; Ali el-Gritly, *The Structure of Modern Industry in* Egypt (Cairo: Government Press, 1948), 432–45, 554–69; and "History of the Customs Regime in Egypt," in *Annual Statement of the Foreign Trade*, Ministry of Finance, Statistical Department (Cairo: Government Press, 1934), 5–13.
10. As the following figures indicate, the percentage of cinemas showing at least some foreign films rose considerably in the postwar years. This pattern held true throughout the country, with the notable exception of Port Said, where an unusual segregation of cinemas—probably due to the lingering British military presence—left only one of fourteen theaters offering a mixed program in 1954.

Year	Foreign %		Egyptian %		Mixed %		Total
1946	58	30	73	38	62	32	193
1952	64	20	77	25	170	55	311
1954	69	19	21	6	264	75	354

Source: Calculated from cinema descriptions in Jacques Pascal, ed., *The Middle East Motion Picture Almanac*, 1st ed. (Cairo: S.O.P. Press, 1946–1947), 19–38; Pascal, *Annuaire du Cinéma pour le Moyen Orient et l'Afrique du Nord* (Cairo: Imprimerie La Patrie, *1951–52*), 49–88; and Pascal, *Annuaire du Cinéma pour le Moyen Orient et l'Afrique du Nord* (Cairo: Imprimerie La Patrie, 1954), 69–119.

11. Mohammed el-Qassass, "Theatre and Cinema," in *Cultural Life in the United Arab Republic*, ed. Mustafa Habib (Cairo: UAR National Commission for UNESCO, 1968), 246.
12. Ahmad Kamel Mursi and Magdi Wahba, *Mu'gam al-fann al-sinima'i* [*A Dictionary of the Cinematic Art*] (Cairo: General Egyptian Book Organization, 1973), 72. Official recognition was granted by Ministerial Decree No. 458 of September 1947. *Cine Film*, no. 14 (January 24, 1949): 6.
13. *Sina'at al-sinima: haqa'iq w-arqam* [*The Film Industry: Facts and Figures*], March 1964, 3. Cinema workers were represented by the Filmmakers' Union (*Niqabat al-Sinima'iyin al-Muhtarafiyin*), which held its first meeting in November 1943. *Mu'gam al-fann al-sinima'i*, 79; Magdi Wahba, *Cultural Policy in Egypt* (Paris: UNESCO, 1972), 60. On the Ministry of Commerce and Industry's forbidding of industrial activity without membership in the appropriate Chamber of the Federation of Egyptian Industries, see *Cine Film*, no. 68 (January 1, 1954): 12.
14. A plan to create an actual monopoly reportedly was developed by the Misr Company for the Theater and Cinema, which sought to unify postwar Egypt's many small producers into one or two companies, with Studio Misr controlling all production. See *Cine Film*, no. 1 (May 1, 1948): 12. Other mergers among exhibitors and producers were rumored in the trade press but failed to materialize more often than not. See *Cine Film*, no. 3 (June 1, 1948): 4. For more on Egyptian corporatism in other areas, see Robert Bianchi, *Unruly Corporatism: Associational Life in Twentieth-Century Egypt* (New York: Oxford University Press, 1989).
15. The cost and earnings figures come from Samir Farid, "Periodization of Egyptian Cinema," in *Screens of Life: Critical Film Writing from the Arab World*, ed. and trans. Alia Arasoughly, 1–18 (Quebec: World Heritage Press, 1996), 8. He provides no source, but these figures are consistent with other estimates, such as those found in *Cine Film*.
16. Red Kann ed., *Motion Picture Almanac, 1951–52* (New York: Quigley, 1952).
17. On the Cinema Chamber's report and local reaction, see *Cine Film*, no. 82 (March 1, 1955): 1.
18. Just after World War II, large Cairo cinemas like the Miami, Royal, Kursal, and Diana closed temporarily in the summer of 1948. *Cine Film*, no. 3 (July 1, 1948):

4. On the overabundance of cinemas, see the lead editorial in *Cine Film*, no. 6 (September 1, 1948): 1.

19. The Federation of Egyptian Industries, *Annuaire, 1955–56* lists the names of the 114 producers and distributors that were members of the Cinema Chamber. The 1958–1959 edition lists only seventy-four, indicating a decline that occurred after the middle of the decade. These figures are for registered, but not necessarily active, firms in the industry.

20. One longtime observer described production as being "at the craftsmanship stage" in this period. While an exaggeration, the small-scale and fleeting nature of many film companies was a defining characteristic of the period. Mazzaoui, *Cinema in the UAR*, 22. For more on industry problems in the mid-1950s, see the Cinema Chamber's Annual Report for 1954, reproduced in *Cine Film*, special issue, no. 89 (October 1955): 29–33.

21. Wahba, *Cultural Policy in Egypt*, 60. Post-1952 union activity included strikes, for example, which raised costs to producers. Law No. 142 of 1955 reorganized the Filmmakers Union, changing it from a labor union to a professional union under the name Union of Cinematic Professions (*Niqabat al-Mihan al-Sinima'iyya*) and creating several branches, along with a Federation of Unions in the professions of theater, cinema, and music. For details, see the law as reprinted in *Cine Film*, special issue, no. 89 (October 1955): 18–27.

22. A breakdown of revenues in this period is in the Cinema Chamber's Annual Report, reprinted in *Cine Film*, special issue, no. 89 (October 1955): 29–33. From it, one can calculate the following for the Egyptian industry in 1954:

Estimated total industry receipts (£E):	
Egyptian films	2,131,500
Foreign films	*913,000*
Total	3,045,000
Entertainment tax (£E):	
Egyptian films	339,469
Gross Profits (£E):	
Exhibitors	896,015 (50% receipts minus tax)
Distributors	89,602 (10% of the above 50%)
Producers	806,413 (40% of the above 50%)
Production costs (£E):	
Total	1,476,000
Average	18,000
Desired returns	1,033,200 (70% of prod. costs)
Actual returns	806,414

23. *Cine Film*, no. 83 (April 1, 1955): 1.

24. This vulnerability was quite literal when the January 1952 riots damaged thirty-two cinemas in Cairo, Alexandria, and the Canal Zone, along with Kodak facilities, film labs, and the J. Arthur Rank Organization office. The latter closed

permanently and sold its theater holdings as a result; many of the damaged cinemas remained closed for months. *Cine Film*, no. 46 (March 1, 1952): 1–8.

25. *Cine Film*, no. 106 (July–August 1957). Taxes included the entertainment tax, municipal tax, charity tax, cinema support tax, and tuberculosis tax, with additional fees for stamps, permits, and the fulfillment of other bureaucratic requirements. Under this much-criticized system, similar theaters in the same neighborhood could pay different amounts of tax for the same film, and large theaters could pay a lower rate than small theaters, all depending on the kinds of tickets bought by customers. For an example of the receipts and taxes for one particular film that played at twenty-four cinemas throughout the country, see *Cine Film*, no. 123 (March 15, 1959): 1–2.
26. The largest chain was owned by Elias Georges Loutfi, who by 1958 had acquired twenty-six cinemas in Cairo and Alexandria. Probably the second largest was controlled by the Cinema Company of the East (*Sharikat al-Sharq lil-Sinima*), which was involved in an ongoing conflict with producers, distributors, and small independent exhibitors. For more on the latter, see the Cinema Chamber meeting notes recorded in *Cine Film*, no. 54 (October 1, 1953): 15.
27. *Cine Film*, no. 103 (January–February 1957): 1; *Cine Film*, no. 60 (May 1, 1953): 1; and the Cinema Chamber's annual report for 1952, reprinted in *Cine Film*, no. 66 (November 1953): 34–35.
28. While never adopted, the proposed quota for the top fifteen distributors reveals the most powerful firms in the period: Benha Films (10 films), Soc. el-Nil (10), Orient Films (10), Dollar Film (6), Hussein Sedky (4), Lotus Films (4), Naguib Nasr (3), Nahas Films (3), Sakr Films (2), Amir Films (2), Hassan Ramzy (2), Farid el-Attrache (2), Hussein Fawzy (2), P. Mouradian (2), and Goumhouriet Masr (2). *Cine Film*, no. 81 (February 1, 1955): 7.
29. "Egypt," *Motion Picture Almanac*, 1956.
30. See Ministerial Decree No. 658 of 1957, as reprinted in Samir Farid, ed., *al-Tashri'at al sinima'iyya fil-watan al-'arabi* [Arab Cinema Legislation] (Cairo: General Union for Arab Artists, 1991), 236. The latter volume contains most of Egypt's major film legislation.
31. Some of the nationalized companies included the distributors Benha Film, Dollar Films, and Orient Films. Mazzaoui, *Cinema in the UAR*, 26.
32. A June 6, 1957, cable from the U.S. Embassy describes how, despite the expiration of a Paramount representative's import license, he was able to bring in five new films and clear them through customs and the censor. See U.S. Department of State, Confidential U.S. State Department Central Files 1955–1959, Record Group 59, Reel 7, Decimal Number 874.452/6-657.
33. For more on the public sector, see Durria Sharaf al-Din, *al-Siyasa wal-sinima fi Misr, 1961–1981* [*Politics and the Cinema in Egypt, 1961–1981*] (Cairo: Dar al-Shuruq, 1992); Muhammed el-Ashari, "Iqtisadiyat sina'at al-sinima fi Misr: dirasa muqarana," ["The Economics of the Film Industry: A Comparative Study"] (PhD diss., Faculty of Law, Cairo University, 1968); and Mohamed Bayoumi, "Intervention de l'Etat dans le cinéma: Problèmes posés par la nationalisation du cinéma en Egypte" ["State Intervention in the Cinema: Problems

Created by the Nationalization of the Cinema in Egypt"] (PhD diss., Université de Paris I, Panthéon-Sorbonne, 1983).

34. The following table indicates the relative size of the public exhibition sector:

Year	Public Cinemas	%	Private Cinemas	%	Total
1967	58	22	205	78	263
1968	54	21	206	79	260
1969	53	22	193	78	246
1970	48	20	190	80	238
1971	49	19	192	80	241
1972	50	20	199	80	249
1973	52	22	187	78	239

Source: Central Agency for Public Mobilization and Statistics, *al-Sinima wal-masrah*, various years.

35. Television was not the only cause of the decline in rural cinemas, since economic policies like import substitution in Nasserist Egypt penalized rural areas and undercut this population's income available for leisure activities.

36. Medhat Mahfouz, "Les salles de projections dans l'industrie cinématographique," in *Egypte: 100 ans de cinema*, ed., Magda Wassef (Paris: Editions Plume and Institute du Monde Arabe, 1995), 126. Average attendance rates were much higher in the 1950s, with a reported frequency of twenty times per year in Cairo and Alexandria in 1954. See *Cine Film*, no. 75 (August 1, 1954): 2.

37. The following figures depict the sources of films exhibited in Egypt for various years:

Year	Foreign %		Egyptian %		Mixed %		Total
1946	58	30	73	38	62	32	193
1952	64	20	77	25	170	55	311
1954	69	19	21	6	264	75	354
1965	63	21	173	58	60	20	296

Source: Calculated from cinema descriptions in Pascal, *The Middle East Motion Picture Almanac*, 19–38; Pascal, *Annuaire du Cinéma*, 49–88; Pascal, *Annuaire du Cinéma*, 69–119; and Ministry of Culture, *Ba'ath: Dur al-Sinima fil-Gumhuriyya al-'Arabiyya al-Mutahida* [*Research Report: The Role of the Cinema in the United Arab Republic*], April 1965, p. 27.

38. For a critique of the revolutionary cinema of the Nasser era, see Raymond William Baker, "Egypt in Shadows: Films and the Political Order," *American Behavioral Scientist* 17, no. 3 (January–February 1974): 393–423.

39. Madkour Thabet, *Egyptian Film Industry* (Cairo: Ministry of Culture, 1998), 25–26.

40. Ministerial decision No. 837 of 1971 set the import quota of 300. Federation of Egyptian Industries, *Yearbook 1973*, 234.

41. See the list of films and their producers in *Sijill al-thaqafa* [*Cultural Registry*] (Cairo: Ministry of Culture, 1987), 256–60.

42. The tax discrimination ratio changed as follows over the years:

48% foreign to 20% local—2.4 to 1 (1989)

48% foreign to 16% local—3 to 1 (1991)

42% foreign to 15% local—2.8 to 1 (1994)

40% foreign to 20% local—2 to 1 (1996)

40% foreign to 20% local—2 to 1 (1998)

Source: Motion Picture Association, "Trade Barriers Report," various years.

43. Samir Farid, personal interview, December 21, 1997.
44. Motion Picture Association, "Trade Barriers Report," various years from 1991 to 1998; and MPA, "Market Background—Egypt," 1989. A list of the many taxes after 1975 is in the Federation of Egyptian Industries, *Yearbook 1975*, 244–45: alms and cinema consolidation duties; fiscal duties on contracts; charges on film production permits; taxes for health purposes; municipal taxes; national security taxes, and so on. See also the list of 14 different taxes in Federation of Egyptian Industries, *Yearbook 1981*, 547–48.
45. There were 120 cinemas in Egypt in 1939. *Cine-Orient*, no. 15 (March 1949): 22.
46. Mahfouz, "Les salles de projections dans l'industrie cinématographique," 128. As Mahfouz points out, the first modern theater to be built outside of the downtown area was the Cinema al-Tahrir in Doqqi, built in 1988.
47. According to one of its chief accountants, Hanna Youssef Hanna, Cairo's landmark Cinema Metro was to be converted into a set of pricey, renovated smaller theaters. Personal interview, November 30, 1997.
48. MPA, "Trade Barriers Report," 1991. Net film rentals are gross receipts minus expenses.
49. Motion Picture Association, "Trade Barriers Report," various years in the 1980s and 1990s.
50. Interview, General Abd el-Rahman, June 17, 1998.
51. See Geoffrey Garrett and Peter Lange, "Internationalization, Institutions, and Political Change," in *Internationalization and Domestic Politics*, ed. Robert O. Keohane and Helen V. Milner, 48–75 (Cambridge: Cambridge University Press, 1996).
52. Charles Ramírez Berg, *Cinema of Solitude: A Critical Study of Mexican Film, 1967–1983* (Austin: University of Texas Press, 1992), 13.
53. *Historia documental del cine mexicano* [*Documental History of the Mexican Cinema*] (Guadalajara: University of Guadalajara; Mexico City: National Council for Culture and the Arts, 1992), 1:11. Cited hereafter as *Historia documental*, followed by the volume and page numbers. This seventeen-volume collection by Mexico's leading film historian reproduces important materials from throughout Mexican film history. The number of producers is calculated from *Historia documental*, various pages.
54. *Historia documental* 1:119.

55. Alexandra Pineda and Antonio Paranaguá, "Mexico and its Cinema," in *Mexican Cinema*, ed. Paulo Antonio Paranaguá, 15–62 (London: British Film Institute, 1995), 24.
56. *Historia documental* 1:211, 253.
57. *Historia documental* 1:120, 2:89.
58. Union struggles are noted in *Historia documental* 2:8, 91. There was, however, a dramatic growth in the number of UTECM (*Unión de Trabajadores de Estudios Cinematográficos de México*) members—from 91 in 1934 to 410 four years later—exceeding the growth of production. *Historia documental* 2:89.
59. *Historia documental* 2:91, 145–46. UTECM was formed in 1936 as part of the STIC. *Historia documental* 1:120.
60. *Historia documental* 2:7, 181, 236.
61. *Historia documental* 3:9.
62. *Historia documental* 3:7–8; 2:181.
63. The producers were Bustillo Oro, Miguel Contreras Torres, De Fuentes, De la Serna, Saisó Piquer, and Zakarías. *Historia documental* 2:237.
64. *Historia documental* 3:10, 109.
65. *Historia documental* 3:10, 213, 220–22.
66. *Historia documental* 3:220, 4:7.
67. For details on Jenkins, see Carl Mora, *Mexican Cinema: Reflections of a Society, 1896–1980*, rev. ed. (Berkeley: University of California Press, 1989), 76–78. Jenkins was widely disliked in the industry, having reportedly arranged his own kidnapping in 1919.
68. *Historia documental* 3:7, 4:105.
69. These distributors included CLASA Films Mundiales, Filmadora Mexicana, and Producciones Rosas Priego. *Historia documental* 4:107.
70. *Historia documental* 4:188, reproducing an interview with Casto Leal, president of the National Cinema Commission, in *Cartel* (September 25, 1948).
71. The peso was devalued from 4.85 to 8.75 to the dollar. Raw film stock and other expenditures still had to be made in hard currency.
72. *Historia documental* 5:7, 4:188.
73. According to Raúl Anda, a producer at the time, there was a virtual loss of all foreign markets except the United States, Venezuela, and Cuba. *Historia documental* 4:187.
74. *Historia documental* 4:187, 191; 5:7, 11.
75. See *Ley y reglamento de la industria cinematográfica* (Mexico City: Secretaría de Gobernación, 1962); and Virgilio Aunduiza, *Legislación cinematográfica mexicana* (Mexico City: *Filmoteca* of the National Autonomous University of Mexico, 1984).
76. *Historia documental* 6:157. An *amparo* is a uniquely Latin American legal institution, roughly equivalent to a writ of habeas corpus, which is a legal instrument designed to bring a party before the courts to assure due process.
77. As García Riera writes, "los partidarios del intervencionismo de estado derrotaron a los defensores del liberalismo económico clásico." *Historia documental* 6:157.

78. *Historia documental* 6:158.
79. *Historia documental* 7:153.
80. That is, *Películas Nacionales*, *Películas Mexicanas*, and *Cimex*. *Historia documental* 8:8. Also, *Historia documental* 6:157, 7:153.
81. *Historia documental* 7:154, 8:7. *Películas Nacionales* comprised thirty-eight production companies by 1958; *Películas Mexicanas* had twenty by then; *Cimex* consisted of fifty-five companies and five foreign affiliates.
82. *Historia documental* 7:155.
83. The Garduño Plan, as summarized by *México cinema* X, no. 30 is reprinted in *Historia documental* 7:7–9.
84. *Historia documental* 7:154–155, 8:8. See the results of an interview with Garduño in *México Cinema*, vol. 9, no. 10: 55, reprinted in *Historia documental* 8:7–8.
85. *Historia documental* 7:156.
86. *Historia documental* 8:7.
87. *Nacional Financiera* was the state development bank, founded in 1934. *Historia documental* 8:7.
88. *Historia documental* 8:7–8.
89. *Historia documental* 10:8, citing former Cinema Bank Director Federico Heuer's *La industria cinematográfica méxicana* (Mexico City: Policromía, 1964).
90. *Historia documental* 9:7, 9; 10:8.
91. America studios produced seventeen of these; there were three independents and seven coproductions. *Historia documental* 9:153.
92. *Historia documental* 10:13–14, reprinting figures from *México en la cultura* (21 September 1959):

Year	Films Produced	Directors	Average Films per Director	New Directors
1938	57	40	1.42	18
1939	37	25	1.48	8
1940	29	20	1.45	4
1941	37	26	1.42	5
1942	47	36	1.31	10
1943	70	44	1.59	10
1944	75	50	1.50	14
1945	81	43	1.88	1
1946	71	39	1.82	1
1947	58	33	1.76	1
1948	81	39	2.08	1
1949	108	47	2.30	0
1950	122	49	2.49	3
1951	101	43	2.35	0
1952	98	41	2.39	3

Year	Films Produced	Directors	Average Films per Director	New Directors
1953	77	36	2.14	1
1954	112	43	2.60	1
1955	83	34	2.44	2
1956	90	35	2.57	2
1957	94	36	2.61	1
1958	104	35	2.97	0

93. *Historia documental* 12:8–9 reprints the full text of the announcement from the publication *La Afición* (August 30, 1964).
94. *Historia documental* 12:181.
95. *Historia documental* 11:8.
96. *Historia documental* 10:160, 11:7–8.
97. *Historia documental* 11:15.
98. *Historia documental* 11:9.
99. Filmmakers apparently went to Guatemala, for example. When the STPC called for forbidding technicians from traveling abroad to work, the president of Guatemala—Mexico's largest foreign market at the time—is said to have responded with a threatened 400 percent tax on a Mexican film imports. *Historia documental* 11:251.
100. *Historia documental* 13:162.
101. The *Centro Universitario de Estudios Cinematográficos* (CUEC) was established at the National Autonomous University of Mexico (UNAM) in 1963. *Historia documental* 14:13–14.
102. *Historia documental* 15:7.
103. *Historia documental* 15:184.
104. *Historia documental* 15:8.
105. Costa, *La "Apertura" cinematográfica*, 82.
106. *Historia documental* 16:157.
107. They went from earning 164 million pesos in 1971 to 360 million pesos in 1976, according to García Riera, *Historia del cine mexicano*, 323.
108. Costa, *La "Apertura" cinematográfica*, 62.
109. García Riera, *Historia del cine mexicano*, 324; and *Historia documental* 17:229.
110. García Riera, *Historia del cine mexicano*, 324.
111. Marcela Fernández Violante, "Mexican Cinema in the 80s," Panel presentation, Universidad Nacional Autónoma de México—San Antonio, February 16, 1989, cited in Berg, *Cinema of Solitude*, 215. The high-grossing *Titanic* is an example of Mexico's role in this global filmmaking process.
112. *Historia documental* 17:230.
113. García Riera, *Historia del cine mexicano*, 351–52.
114. The three circuits were COTSA, *Cadena Real*, and *Grupo Intercine*. Paranaguá, "Mexico and Its Cinema," in *Mexican Cinema*, 59.

115. Mexico made about 100 films in 1989, just over 50 in 1990, 34 in 1991, and 11 in 1995.

Chapter 5

1. The balance of institutional authority for a given area is both an empirical question—one of historically determined and observable fact—and a political one, reflecting allocative choices made by state authorities.
2. More specifically, institutional parity is measured by dividing into two categories all state institutions with regulative authority over the film sector at a given time. Those using purely economic criteria to regulate are distinguished from those that incorporate legally codified cultural concerns. If cultural interests are not represented in the array of institutions, then a lower level of institutional parity is assigned. If cultural interests are represented, and the state institution using those criteria is administratively autonomous from the state's economic institutions, then a higher level of institutional parity is designated.
3. Lloyd I. Rudolph, "Establishing a Niche for Cultural Policy: An Introduction," in *Cultural Policy in India*, ed. Lloyd I. Rudolph (Delhi: Chanakya Publications, 1984), 2.
4. *Cine Film*, no. 26 (June 1950): 10; *Mu'gam al-fann al-sinima'i*, 70.
5. *International Motion Picture Almanac, 1937–1938* (New York: Quigley, 1938), 1124.
6. For one such articulation, see the U.S. State Department report on Egypt's early production code in Confidential U.S. State Department Central Files (declassified), *Egypt: Internal Affairs and Foreign Affairs*, Record Group 59, Reel 8, Motion Pictures, 883.4061.
7. Magda Wassef, ed., *Egypte: 100 ans de cinéma* (Paris: Editions Plume and Institute du Monde Arabe, 1995), 23.
8. Ibid., 20; *Mu'gam al-fann al-sinima'i*, 70–71.
9. Decree of August 23, 1928, in the *Journal Officiel*, no. 77 (August 30, 1928): reported in *L'Egypte contemporaine*, no. 116–17 (November–December 1929). Egypt's first cinema regulation actually came in 1911 from the Ministry of Interior. Jacques Pascal, ed., *The Middle-East Motion Picture Almanac/al-Dalil al-sinima'i lil-sharq al-awsat*, 1st ed. (Cairo: S.O.P. Press, 1947).
10. Michael W. Albin, "Official Culture and the Role of the Book," *Journal of the American Research Center in Egypt* 24 (1987).
11. See the book-length chronology, Aly Abu Shadi, *Waqa'i' al-sinima al-misriyya fi mi'at 'am: 1896-1995* [100 Years of Developments in the Egyptian Cinema, 1896–1995] (Cairo: al-Hay'at al-'Amma lil-Shu'un al-Mutabi' al-Amiriyya, 1997), 163.
12. On the age restrictions, see Law No. 427 of 1954 in *al-Tashri'at al-sinima'iyya*, 301–302. On the organization of censorship, see Law No. 430 and the Ministry of National Guidance's implementing rules in Ministerial Decree No. 163 of

1955, reprinted in *al-Tashri'at al-sinima'iyya*, 280–85. Also Samir Farid, "La censure, mode d'emploi," in Wassef, ed., *Egypte*, 107.

13. Jack Crabbs Jr., "Politics, History, and Culture in Nasser's Egypt," *International Journal of Middle East Studies* 6, no. 4 (1975): 405. Tellingly, Law No. 372 of 1956 included the cinema in its definition of *malhan*—places of entertainment or amusement—along with night clubs and dance halls. According to film critic Samir Farid, the word "entertainment" (*tasliya*) has a lowbrow connotation in Arabic that, by its application to cinema, made Egyptians think less of the cinema, almost sounding sordid and trivial in a way to which Americans are unaccustomed. To this day, popular Egyptian attitudes toward the cinema are like those of many Mexicans, often displaying a sort of schizophrenia: pride in the historical accomplishments of the industry combined with disdain and some embarrassment over its failures. This has increased in Egypt since the 1970s, when the deteriorating condition of both movie theaters and much of what was shown there led many middle-class moviegoers to stop considering film as a respectable form of entertainment. Long-standing Islamist criticism of the cinema also may have contributed to its poor reputation. Samir Farid, personal interview, December 21, 1997.
14. An essential study of filmmaking under Nasser is Joel Gordon, *Revolutionary Melodrama: Popular Film and Civic Identity in Nasser's Egypt* (Chicago: Middle East Documentation Center, 2002).
15. For more on the Supreme Council (*al-Majlis al-A'ala lil-Ri'aya al-Funun wal-Adab wal-'Ulum al-Ijtima'iyya*) and its establishment by Law No. 4 of 1956, see its Annual Report: *al-Taqrir al-sanawi*, 1958-59 (Cairo: n.p.).
16. Samir Farid, "Periodization of Egyptian Cinema," in *Screens of Life: Critical Film Writing from the Arab World*, ed. and trans. Alia Arasoughly, 1–18 (Quebec: World Heritage Press, 1996), 10.
17. See Law No. 373 of 1956 in Samir Farid, ed., *al-Tashri'at al-sinima'iyya fil-watan al-'arabi* [Arab Cinema Legislation] (Cairo: General Union for Arab Artists, 1991), 211–12.
18. Openings included *Rodda Qalbi* and the Ismail Yassin series of comedies in which the latter plays a young recruit in various Egyptian state institutions like the Army, Navy, Secret Police, and so on. As noted earlier, Nasser hosted Cecil B. DeMille when the latter came to Egypt to film *The Ten Commandments* in 1954.
19. Ministerial Decree No. 60 of 1958, modifying Ministerial Decree No. 658 of 1957, in *al-Tashri'at al-sinima'iyya*.
20. Abu Shadi, *Waqa'i' al-sinima al-misriyya*, 185–86. Abu Shadi describes these politically serious films as "al-intaj al-sinima'i al-hadif," and Samir Farid notes elsewhere that the term "al-hadif" is used to characterize film "with a purpose," that is, socialist "propaganda." See Farid, "Periodization," 12.
21. Abu Shadi, *Waqa'i' al-sinima al-misriyya*, 191.
22. Ali Abu Shadi, "Le secteur public 1963-1972: esquisse d'analyse économique et artistique," Wassef, ed., *Egypte*, 120. See also the statement by Okasha in *Cine Film*, no. 124 (April 16, 1959).

23. Ahmad Kamel Mursi and Magdi Wahba, *Mu'gam al-fann al-sinima'i* [*A Dictionary of the Cinematic Art*] (Cairo: General Egyptian Book Organization, 1973), 76.
24. Presidential Decree No. 48 joined the cinema with the radio and television organization. The motion picture section comprised four companies with a capitalization of £E 2,250,000:

    1. General Company for the Production of Arab Films
    2. General Egyptian Company for International Film Production
    3. General Company for Cinema Studios
    4. General Company for Distribution and Exhibition of Cinematic Films.

    Two additional companies were founded in 1964, the General Company for Cinema Houses and the Cairo Company for Cinema. For more details, see Abu Shadi, *Waqa'i' al-sinima al-misriyya*, 207; Medhat Mahfouz, Les salles de projections dans l'industrie cinématographique," in Wassef, ed., *Egypte*, 127; *Sina'at al-sinima: haqa'iq w-arqam* [*The Film Industry: Facts and Figures*] February 1965, 3.
25. *Sina'at al-sinima: haqa'iq w-arqam*, February 1965, p. 3. The new ministry proved untenable and had its functions returned the next year to other bodies dealing with foreign cultural cooperation.
26. *Sina'at al-sinima: haqa'iq w-arqam*, April 1971, p. 6. This technical center was the Visual Images Technical Center (*al-Markaz al-Fanni lil-Suwwar al-Mar'iyya*). Magdi Wahba, *Cultural Policy in Egypt* (Paris: UNESCO, 1972), 59.
27. The latter was the *Idara al-Thaqafa al-Jamahiriyya. Mu'gam al-fann al sinima'i*, 71.
28. Crabbs, "Politics, History, and Culture in Nasser's Egypt," 386–87.
29. Abu Shadi, *Waqa'i' al-sinima al-misriyya*, 120. See the Minister of Culture's statement: Abdul Kader Hatem, "Cultural Planning and the Features of Our Cultural Revolution," *Egyptian Political Science Review/al-Majalla al-misriyya lil-'ulum al-siyasiyya*, no. 26 (May 1963): 3–15.
30. Abu Shadi's discussion of this is particularly interesting. For a list of all public sector films released from 1963–75, see *al-Sinima wal-tarikh*, no. 2 (1992): 103–9. For a well-known 1981 report on public sector losses, see *al-Sinima wal-tarikh*, no. 7 (1993): 77–91.
31. Decree No. 453 of 1966 replaced the General Egyptian Organization for the Cinema, Radio, and Television with the General Egyptian Organization for the Cinema. Decree No. 48 of 1966 joined the General Company for Cinemas and the General Company for Distribution and Exhibition of Cinematic Films, under the new name Cairo Company for Cinematic Distribution. Decree No. 49 of 1966 joined the General Egyptian Company for World Cinematic Production and the General Company for Cinema Studios and the General Company for Arab Cinematic Production into the Cairo Company for Cinematic Production. Later, by Presidential Decree No. 511 of 1970, the Cairo Company for Cinematic Production and the Cairo Company for Cinematic Distribution were both

folded into the General Egyptian Organization for the Cinema. (*Sina'at al-sinima: haqa'iq w-arqam*, April 1971, pp. 6–7).

32. Budget figures come from Wahba, *Cultural Policy in Egypt*, 27, citing the Ministry of Culture's Department of Research and Planning. Observers differ greatly on the critical success of public sector filmmaking, but Shadi Abd el-Salam's acclaimed film, *el-Mumiya* (*The Night of Counting the Years*), was made late in the period.
33. Abu Shadi, *Waqa'i' al-sinima al-misriyya*, 247.
34. Tharwat Okasha, *al-Siyasa al-thaqafiyya: bayan al-duktur Tharwat Okasha wazir al-thaqafa amam lajnat al-khidamat bi majlis al-umma fi 16 yunyu 1969* [Cultural Policy: Statement of Dr. Tharwat Okasha, Minister of Culture, Before the Services Committee of the National Council, July 16, 1969] (Cairo: Dar al-Kutub, 1969), esp. 5–19 (general cultural policy) and 108–32 (on the cinema). The speech is also discussed in Wahba, *Mu'gam al-fann al-sinima'i*, 34–36.
35. Wahba, *Cultural Policy in Egypt* has a useful organigram of the Ministry of Culture, circa 1970 (89).
36. Wahba, *Cultural Policy in Egypt*, 17. When he wrote this in 1972, Wahba was an Under-Secretary in the Ministry of Culture and a Cairo University professor of English.
37. Presidential Decree No. 2827 of November 1971 joined the General Organization for the Cinema with the General Organization for the Theater, Music and Popular Arts. Federation of Egyptian Industries, *Yearbook 1973*, 230. While the state eliminated the Cinema Organization, it did not stop collecting the cinema support tax at the box office, funneling the latter directly into state coffers. Federation of Egyptian Industries, *Yearbook 1981*, 551.
38. Law No. 13 of 1971 was replaced subsequently by ministerial decision No. 523 of 1973, giving Egyptian movies first priority and requiring the continued projection of any film with revenues equaling £E 1600 per week. Federation of Egyptian Industries, *Yearbook 1974*, 225.
39. See *el-Tashri'at al-sinima'iyya*, Decree No. 181 of 1973, as well as the discussion in Federation of Egyptian Industries, *Yearbook 1974*, 225; and *Yearbook 1975*, 242.
40. *Sina'at al-sinima: haqa'iq w-arqam*, General Egyptian Organization for the Cinema, Radio, and Television, May 1975, p. 4.
41. As former state official Wahba notes, "Next to the secondary school, the house of culture is the most active centre of permanent education in the provinces" (*Cultural Policy in Egypt*, 72).
42. Samir Farid, "La censure, mode d'emploi," in Wassef, ed., *Egypte*, 109. It was not prudishness or piety that drove strict censorship in matters of morality, but the political threat of long-standing conservative Islamist opposition to the cinema.
43. Abu Shadi, *Waqa'i' al-sinima al-misriyya*, 256. Significantly, the Cinema Chamber was not reestablished by the Ministries of Culture or Information, but by Ministerial Decree No. 203 of the Ministry of Industry, Petroleum, and Mining. Federation of Egyptian Industries, *Yearbook 1973*, 230.

44. Producers received five votes, compared with one vote for each of the other subsectors (e.g., distributors and studios). *Mu'gam al-fann al-sinima'i*, 72.
45. On the *munfatihun* ("openers"), see John Waterbury, *The Egypt of Nasser and Sadat: The Political Economy of Two Regimes* (Princeton, NJ: Princeton University Press, 1983), 173–75. Similarly, an entire genre of films developed after 1975, attacking the Nasser-era's "centers of power" *(marakiz al-quwwa)*. See Abu Shadi, *Waqa'i' al-sinima al-misriyya*, 276.
46. Decree No. 22 of the Ministry of Culture and Information. Abu Shadi, *Waqa'i' al-sinima al-misriyya*, 283.
47. These included the Misr Company for Production and Studios and the Misr Company for Distribution and Exhibition, both of which were overseen by the Public Sector Commission for the Cinema and Light Shows (*Hay'at al-Qita'a al-'Am lil-Sinima wal-Daw'iyat*). See the organigram in *Sijill al-thaqafa, 1987–1988*, 166.
48. Abu Shadi, *Waqa'i' al-sinima al-misriyya*, 298.
49. See the various reports issued from 1979 to 1991: *Taqrir al-Majlis al-Qawmi lil-Thaqafa wal-Funun wal-Adab wal-I'lam*, Specialized National Councils, Cairo.
50. *Sijill al-thaqafa*, 1979–80 [Cultural Register], Supreme Council for Culture, General Administration for Planning and Mobilization (Cairo).
51. *Sijill al-thaqafa, 1979–1980*, 23–24.
52. These were the so-called contracts films (*aflam al-muqawalat*).
53. Abu Shadi, *Waqa'i' al-sinima al-misriyya*, 312.
54. See the organigram in *al-Sinima wal-tarikh* [Cinema and History], no. 7, (1993): 66. On the Cultural Development Fund, see Abu Shadi, *Waqa'i' al-sinima al-misriyya*, 325. For an organigram of its various components, see *Sijill al-thaqafa, 1994*, 403. The Cinema Support Fund was absorbed by the Cultural Development Fund in 1989. That same year, the Public Agency for Cinema Palaces (*al-Hay'at al-'Amma li Qusur al-Thaqafa*), which is the institutional legacy of *al-Thaqafa al-Jamahiriyya*, opened its first cinema palace in Garden City, Cairo.
55. Ministerial Decree No. 235 of 1983, in *al-Tashri'at al-sinima'iyya*.
56. Abu Shadi, *Waqa'i' al-sinima al-misriyya*, 302, 316.
57. See Ministry of Culture organigrams in *al-Sinima wal-tarikh*, no. 7 (1993): 66–70; and in *Sijill al-thaqafa* [Cultural Register], *1987–1988*, 166.
58. The £E 100,000 actually went to the film's director, Sherif 'Arafa. In 1993, Imam fronted a series of protests against Islamist violence in Cairo, where free booklets in a series entitled, *al-Muwajiha* (The Confrontation) were distributed, reprinting some of the works of high-profile classic and contemporary secularists, such as Taha Hussein (*The Future of Culture in Egypt*), Qassem Amin (*Liberation of the Woman*), and Gaber 'Asfour, *The Enlightenment Confronts Oppression*. For more on *Terrorism and Kebab*, see Raymond Baker, "Combative Cultural Politics: Film Art and Political Spaces in Egypt," *Alif* 15 (1995).
59. One of Egypt's best-known filmmakers, Youssef Chahine, was sued by the Islamist opposition in 1994 for his film *al-Muhager*.
60. Aurelio de los Reyes, "The Silent Cinema," in *Mexican Cinema, ed.* Paulo Antonio Paranaguá, 63–78 (London: British Film Institute, 1995), 67, 71.

61. For general coverage, see Gilbert M. Joseph, Anne Rubenstein, and Eric Zolov, eds. *Fragments of a Golden Age: The Politics of Culture in Mexico Since 1940* (Durham, NC: Duke University Press, 2001).
62. Aurelio de los Reyes, "The Silent Cinema," 76.
63. *Historia documental del cine mexicano* [*Documental History of the Mexican Cinema*], *1938–1942*, (Guadalajara and Mexico City: University of Guadalajara and National Council for Culture and the Arts, 1992), 1:13, 31. Cited hereafter as *Historia documental*, followed by the volume and page numbers.
64. *Historia documental* 1:76.
65. CLASA was actually run by a former government official who had been, among other things, head of the Departments of Industry and the Treasury. *Historia documental* 1:163.
66. *Historia documental* 1:120, 213.
67. *Historia documental* 2:89, 146.
68. *México Cinema*, 34.
69. *Historia documental* 2:146.
70. *Historia documental* 2:182.
71. This image of "Mexicanness" consisted of respect and politeness, deference to higher classes, sentimentality, and attentiveness to courtesy and manners. Prof. Mariano Sánchez-Ventura, UNAM, July 8, 1996.
72. *Historia documental* 1:76.
73. *Historia documental* 2:182; 4:12.
74. *Historia documental* 2:111, citing *Cine mexicano* 16, no. 12:44.
75. *Historia documental* 3:111; 4:107, 187.
76. The law is reprinted in full in *Historia documental* 5, 11–14.
77. See Carlos Monsiváis, "Mexican Cinema: Of Myths and Demystifications," in *Mediating Two Worlds: Cinematic Encounters in the Americas*, ed. John King, Ana M. López, and Manuel Alvarado, 139–46 (London: British Film Institute, 1993).
78. *Ley y reglamento de la industria cinematográfica* (Mexico City: Secretaría de Gobernación, 1962).
79. *Historia documental* 5:13. Producers and exhibitors of Mexican films were given two representatives each. Other members included the Department of Interior (president), the Department of Foreign Relations, the Department of Treasury and Public Credit, the Ministry of the Economy, the Department of Public Education, the Department of the Federal District, the General Film Office, the National Cinema Bank, the studios, labs, exhibitors, the STPC union, and the STIC union.
80. *Historia documental* 6:158. The 1952 law gave the Interior Department control over any changes in the construction, purchase, or use of studios and film labs, as well as changes in existing equipment or the import of such equipment. *Historia documental* 7:9.
81. *Historia documental* 11:132.
82. *Historia documental* 5:171, citing Raúl Cremoux, *La televisión y el alumno de secundaria del Distrito Federal.*

83. Part of the message is reprinted in *Historia documental* 6:8.
84. Uruchurtu's speech is reprinted in *Historia documental* 6:8.
85. *Historia documental* 6:158–59.
86. The head of the Cinema Bank, Eduardo Garduño, used the term "interés nacional." See the 1955 interview in *México Cinema* 10, 9 (1955), reprinted in *Historia documental* 7:8.
87. As a percentage of its production, Mexico reportedly produced more color films in 1956 than any other country in the world. *Historia documental* 8:7, 169.
88. *Historia documental* 7:154, 9:157.
89. *Historia documental* 10:156.
90. *Historia documental* 10:158–60.
91. *Historia documental* 9:153–54; 10:157–59.
92. *Historia documental* 10:157; and Carl J. Mora, *Mexican Cinema: Reflections of a Society*, rev. ed. (Berkeley: University of California Press, 1989), 101.
93. The film *Rosa Blanca*, for example, dealt with oil companies' expropriation of agricultural land. It was banned even though it had been produced with partial state funding, reportedly out of a desire to avoid upsetting the U.S. companies involved. *Historia documental* 11:21.
94. Argentine producers had incentives to make higher-quality films, while Mexican filmmakers were forced by the incentive structure to pursue purely commercial objectives.
95. *Historia documental* 11:131, 138.
96. The announcement is reprinted in *Historia documental* 12:8–9.
97. *Historia documental* 13:164; and 14:195–96.
98. Heuer, *La Industria cinematográfica mexicana*, 179–84.
99. *Historia documental* 11:14, from Salvador Elizondo, "El cine mexicano y la crisis," *Nuevo Cine* 7 (1962).
100. *Historia documental* 12:8–9, 14:10, on changes in the DGC.
101. *Historia documental* 15:184.
102. *Historia documental* 17:7, citing Paola Costa, *La 'Apertura' cinematográfica: México, 1970–1976* (Puebla: University of Puebla, 1988), 74.
103. *Historia documental* 17:7, citing Costa, *La 'Apertura' cinematográfica*, 75.
104. *Historia documental* 15:8–9, 186; 17:9–10, citing Costa, *La 'Apertura' cinematográfica*, 82.
105. *Historia documental* 17:10. Legislation creating the film school, the *Centro de Capacitación Cinematográfica*, had been passed several years earlier, but not until 1975 was it actually established. *Historia documental* 17:119.
106. The following figures show the rise and decline of state involvement in production throughout the 1970s.

Year	State Productions	Total Productions	Percentage
1970	0	95	0%
1971	2	88	2.3%
1972	16	90	18%
1973	16	73	22%

Year	State Productions	Total Productions	Percentage
1974	20	67	30%
1975	23	60	38%
1976	36	60	60%
1977	44	77	57%
1978	29	107	27%
1979	15	113	13%
1980	5	107	5%
1981	7	97	7%
1982	7	87	8%

Source: *Historia documental*, various years and *Historia del cine mexicano*, 323.

107. *Historia documental* 17:229.
108. Emilio García Riera, *Historia del cine mexicano* (Mexico City: Secretaría de Educacíon Pública, 1986), 324. The final liquidation of the Cinema Bank did not occur until the de la Madrid administration several years later.
109. *Historia documental* 17:229–30. The old peso was worth 1/1000 of the post-1993 peso.
110. García Riera, *Historia del cine mexicano*, 323–25; and Tomás Pérez Turrent, "Crises and Renovation," in *Mexican Cinema*, ed. Paulo Antonio Paranaguá, 94–116 (London: British Film Institute and IMCINE), 108.
111. García Riera, *Historia del cine mexicano*, 351.
112. Tomás Pérez Turrent, who is paraphrasing Antonio Bardem's description of the Spanish cinema of the 1950s, as quoted in Mora, *Mexican Cinema*, 147. The fuller description is that Mexican cinema is "[c]heap cinema, with no artistic or cultural ambitions, a repetition of the oldest and most time-worn formulas, subpornography, facile folklore, routine melodramas, films aimed at manipulating the emotions and frustrations of the lower strata of the population and the nostalgia of Mexicans and those of Mexican descent in the United States."
113. García Riera, *Historia del cine mexicano*, 351.
114. "Se Dinamiza la Cineteca Nacional," *Hispanoamericano* (July 28, 1987), 65, as cited and translated in Mora, *Mexican Cinema*, 153.
115. Teresa Zacarias Figueroa, Assistant Director of Diffusion and International Events, IMCINE, personal interview, July 25, 1996.
116. Bernardo Stril, Assistant Director of Promotion, IMCINE, personal interview, July 30, 1996, at IMCINE, Mexico City. IMCINE's promotional material lists several ambitious objectives for its 1995–2000 program, including promoting Mexico's "artistic and cultural heritage," improving the quality of films, advancing cultural and democratic pluralism, disseminating Mexican culture, as well as furthering industry deregulation and meeting global competition.
117. Bernardo Stril, July 30, 1996.

118. IMCINE's promotional material and the Assistant Director for Promotion both claim the agency may fund up to 60 percent of select joint ventures, but such a high percentage is probably very rare.
119. This is according to filmmaker Alejandro Pelayo, cited in Mora, *Mexican Cinema*, 183. Some of the films IMCINE supported in the 1990s included Jorge Fons's *El Callejón de los Millagros*, an adaptation of Egyptian novelist Naguib Mahfouz's *Midaq Alley*; María Novaro's critically acclaimed, *Danzon*; Alfonso Arau's successful film, *Como Agua para Chocolate* [*Like Water for Chocolate*]; and the late Tomás Gutiérrez Alea's Cuban-Mexican-Spanish coproduction, *Fresa y Chocolate* [*Strawberry and Chocolate*].
120. Mario Aguiñaga-Ortuño, Director of the *Cineteca Nacional*, personal interview, August 5, 1996.

Chapter 6

1. See Robert Gilpin, *U.S. Power and the Multinational Corporation* (New York: Basic Books, 1975).
2. The 1996 Federal Communications Act, the first revamping of U.S. communications law in sixty years, went even further to deregulate telecommunications, reducing restrictions on the use of public radio and television for political advertisements on the assumption that the market mechanism would regulate them effectively. Massive media consolidation since 1996 has called into question the legislation's effectiveness.
3. See Peter Haas, "Do Regimes Matter? Epistemic Communities and Mediterranean Pollution Control," *International Organization* 43, no. 3 (Summer 1989): 377–403.
4. For a related argument, see Hendrik Spruyt, *The Sovereign State and Its Competitors: An Analysis of Systems Change* (Princeton, NJ: Princeton University Press, 1994).
5. Similar kinds of claims are developed by the late Stephen Jay Gould in, for example, "The Panda's Thumb of Technology," in *Bully for Brontosaurus: Reflections in Natural History*, ed. Gould, 59–75 (New York: W. W. Norton, 1991).
6. Ronald Dore, "Convergence in Whose Interest?" in *National Diversity and Global Capitalism, ed.* Suzanne Berger and Ronald Dore, 366–74 (Ithaca, NY: Cornell University Press, 1996), 367.
7. For Adam Smith's discussion of education and "public diversions," see *The Wealth of Nations*, vol. 2 (New York: G. P. Putnam's Sons, 1904), 269–70, 281. As one remedy to moral decay, he writes, "The state, by encouraging, that is by giving entire liberty to all those who for their own interest would attempt, without scandal or indecency, to amuse and divert the people by painting, poetry, music, dancing; by all sorts of dramatic representations and exhibitions, would easily dissipate, in the greater part of them, that melancholy and gloomy humour which is almost always the nurse of popular superstition and enthusiasm" (281).

8. The role of censorship in literature, for example, was seen in Czarist Russia, where heavy-handed state censorship actually aided the artistic development of the Russian novel in the late nineteenth century. For an insightful historical account of the complex relationship between repressive state power and cultural production, see Marla Susan Stone, *The Patron State: Culture and Politics in Fascist Italy* (Princeton, NJ: Princeton University Press, 1998).
9. How would most Americans feel if, for example, the vast majority of the films at local cinemas were in Chinese, or if international news reports centered on Tokyo, fashion emanated from Seoul, or Beijing produced most of the world's ideas—or ideals?

Bibliography

Books

Abu Shadi, Aly. *Waqa'i' al-sinima al-misriyya fi mi'at 'am: 1896–1995*. Cairo: al-Hay'at al-'Amma lil-Shu'un al-Mutabi' al-Amiriyya, 1997.

———. "Chronologie: 1896–1994." In *Egypte: 100 ans de cinema*. Edited by Magda Wassef, 18–39. Paris: Editions Plume and Institute du Monde Arabe, 1995.

Adorno, Theodor. *The Culture Industry: Selected Essays on Mass Culture*, edited by J. M. Bernstein. London: Routledge, 1991.

Anderson, Benedict. *Imagined Communities: Reflections on the Origins and Spread of Nationalism*. Rev. ed. New York: Verso, 1991.

Andrew, Dudley. *Mists of Regret: Culture and Sensibility in Classic French Film*. Princeton, NJ: Princeton University Press, 1995.

Appadurai, Arjun. *Modernity at Large: Cultural Dimensions of Globalization*. Minneapolis: University of Minnesota Press, 1996.

Arasoughly, Alia, ed. and trans. *Screens of Life: Critical Film Writing from the Arab World*. Vol. 1. Quebec: World Heritage Press, 1996.

Armbrust, Walter. *Mass Culture and Modernism in Egypt*. Cambridge: Cambridge University Press, 1996.

———, ed. *Mass Mediations: New Approaches to Popular Culture in the Middle East and Beyond*. Berkeley: University of California Press, 2000.

Armes, Roy. *Third World Film Making and the West*. Berkeley: University of California Press, 1987.

el-Ashari, Muhammed. "Iqtisadiyat sina'at al-sinima fi Misr: dirasa muqarana." PhD diss., Cairo University, 1968.

Aunduiza, Virgilio. *Legislación cinematográfica mexicana*. Mexico City: National Autonomous University of Mexico, 1984.

Ayad, Christophe. "Le star-système: De la splendeur au voile." In *Egypte: 100 ans de cinema*. Edited by Magda Wassef, 134–41. Paris: Editions Plume and Institute du Monde Arabe, 1995.

Badrakhan, Ahmad. *al-Sinima*. Cairo: al-Halabi Press, 1936.

Balio, Tino, ed. *The American Film Industry*. Rev. ed. Madison: University of Wisconsin Press, 1985.

Barnouw, Erik, and S. Krishnaswamy. *Indian Film*. New York: Oxford University Press, 1980.

Bayoumi, M. "Intervention de l'Etat dans le cinéma: Problèmes posés par la nationalization du cinéma en Egypte." PhD diss., Université de Paris I; Panthéon-Sorbonne, 1983.

Belassa, Bela. *The Structure of Protectionism in Developing Countries*. Baltimore: Johns Hopkins University Press, 1971.

Berg, Charles Ramírez. *Cinema of Solitude: A Critical Study of Mexican Film, 1967–1983*. Austin: University of Texas Press, 1992.

Berger, Suzanne, and Ronald Dore, eds. *National Diversity and Global Capitalism*. Ithaca, NY: Cornell University Press, 1996.

Bianchi, Robert. *Unruly Corporatism: Associational Life in Twentieth-Century Egypt*. Oxford: Oxford University Press, 1989.

Bordwell, David, Janet Staiger, and Kristin Thompson. *The Classical Hollywood Cinema: Film Style and Mode of Production to 1960*. New York: Columbia University Press, 1985.

Borneman, Ernest. "United States versus Hollywood: The Case Study of an Antitrust Suit." In *The American Film Industry*. Rev. ed. Edited by Tino Balio, 449–62. Madison: University of Wisconsin Press, 1985.

Bourdieu, Pierre. *The Field of Cultural Production: Essays on Art and Literature*. Cambridge: Polity Press, 1993.

Camp, Roderic A. *Intellectuals and the State in Twentieth-Century Mexico*. Austin: University of Texas Press, 1985.

Cardoso, Fernando Henrique, and Enzo Faletto. *Dependency and Development in Latin America*. Berkeley: University of California Press, 1979.

Carlton, Dennis W., and Jeffrey M. Perloff. *Modern Industrial Organization*. 2nd ed. New York: HarperCollins, 1994.

Caves, Richard E. *Creative Industries: Contracts Between Art and Commerce*. Cambridge, MA: Harvard University Press, 2002.

Chanan, Michael. *Labour Power in the British Film Industry*. London: British Film Institute, 1976.

Chandler, Alfred D., Jr. "The Evolution of Modern Global Competition." In *Competition in Global Industries*. Edited by Michael E. Porter, 405–48. Boston: Harvard Business School Press, 1986.

Conant, Michael. *Antitrust in the Motion Picture Industry*. Berkeley: University of California Press, 1960.

———. "The Paramount Decrees Reconsidered." In *The American Film Industry*. Rev. ed. Edited by Tino Balio, 537–73. Madison: University of Wisconsin Press, 1985.

Costa, Paola. *La "Apertura" cinematográfica: México, 1970–1976*. Puebla, Mexico: Universidad Autónoma de Puebla, 1983.

Cowen, Tyler. *In Praise of Commercial Culture*. Cambridge, MA: Harvard University Press, 1998.

Cummings, Milton C., Jr., and Richard S. Katz, eds. *The Patron State: Government and the Arts in Europe, North America, and Japan*. New York: Oxford University Press, 1987.

Cvetkovich, Ann, and Douglas Kellner, eds. *Articulating the Global and the Local: Globalization and Cultural Studies*. Boulder, CO: Westview, 1997.

Dajani, Karen F. "Egypt's Role as a Major Media Producer, Supplier and Distributor to the Arab World: An Historical–Descriptive Study." PhD diss., Temple University, 1979.

Danan, Martine. "From Nationalism to Globalization: France's Challenges to Hollywood's Hegemony." PhD diss., Michigan Technological University, 1994.

Darwish, Mustafa. *Dream Makers on the Nile: A Portrait of Egyptian Cinema*. Cairo: American University in Cairo Press, 1998.

Davis, Eric. *Challenging Colonialism: Bank Misr and Egyptian Industrialization*. Princeton, NJ: Princeton University Press, 1982.

de Usabel, Gaizka. *The High Noon of American Films in Latin America*. Ann Arbor, MI: UMI Research Press, 1982.

Doherty, Thomas. *Projections of War: Hollywood, American Culture, and World War II*. New York: Columbia University Press, 1993.

Dorfman, Ariel, and Armand Mattelart. *How to Read Donald Duck: Imperialist Ideology in the Disney Comic*. Trans. and rev. ed. New York: International General, 1991.

Evans, Peter. *Dependent Development: The Alliance of Multinational, State, and Local Capital in Brazil*. Princeton, NJ: Princeton University Press, 1979.

———. *Embedded Autonomy: States and Industrial Transformation*. Princeton, NJ: Princeton University Press, 1995.

Farid, Samir. "Periodization of Egyptian Cinema." In *Screens of Life: Critical Film Writing from the Arab World*. Vol. 1. Edited and translated by Alia Arasoughly, 1–18. Quebec: World Heritage Press, 1996.

———, ed. *al-Tashri'at al-sinima'iyya fil-watan al-'arabi*. Cairo: General Union for Arab Artists, 1991.

Fielding, Raymond, ed. *A Technological History of Motion Pictures and Television. An Anthology from the Journal of the Society of Motion Picture and Television Engineers*. Berkeley: University of California Press, 1967.

Finler, Joel W. *The Hollywood Story*. New York: Crown, 1988.

Frieden, Jeffry. *Debt, Development, and Democracy: Modern Political Economy and Latin America, 1965–1985*. Princeton, NJ: Princeton University Press, 1991.

Fulton, A. R. "The Machine." In *The American Film Industry*. Rev. ed. Edited by Tino Balio, 27–42. Madison: University of Wisconsin Press, 1985.

García Riera, Emilio. *Historia del cine mexicano*. Mexico City: Secretaría de Educacíon Pública, 1986.

———. *Historia documental del cine mexicano*. Vols. 1–17. Guadalajara: University of Guadalajara; Mexico City: National Council for Culture and the Arts, 1992.

Garrett, Geoffrey. *Partisan Politics in the Global Economy*. Cambridge: Cambridge University Press, 1998.

Garrett, Geoffrey, and Peter Lange. "Internationalization, Institutions, and Political Change." In *Internationalization and Domestic Politics*. Edited by Robert O. Keohane and Helen V. Milner, 48–75. Cambridge: Cambridge University Press, 1996.

Gilpin, Robert. *War and Change in World Politics*. Cambridge: Cambridge University Press, 1981.

———. *U.S. Power and the Multinational Corporation*. New York: Basic Books, 1975.

Golding, Peter, and Graham Murdock. "Culture, Communications, and Political Economy." In *Mass Media and Society*. 2nd ed. Edited by James Curran and Michael Gurevitch, 15–32. London: Arnold, 1996.

Gomery, Douglas. "The Coming of Sound to the American Cinema: A History of the Transformation of an Industry." PhD diss., University of Wisconsin at Madison, 1975.

Gordon, Joel. *Revolutionary Melodrama: Popular Film and Civic Identity in Nasser's Egypt*. Chicago: Middle East Documentation Center, 2002.

Gould, Stephen Jay. "The Panda's Thumb of Technology." In *Bully for Brontosaurus: Reflections in Natural History*, 59–75. New York: W. W. Norton, 1991.

Gourevitch, Peter. *Politics in Hard Times: Comparative Responses to International Economic Crises*. Ithaca, NY: Cornell University Press, 1986.

el-Gritly, Ali. *The Structure of Modern Industry in Egypt*. Cairo: Government Press, 1948.

Guback, Thomas. *The International Film Industry: Western Europe and America Since 1945*. Bloomington: Indiana University Press, 1969.

———. "Shaping the Film Business in Postwar Germany: The Role of the U.S. Film Industry and the U.S. State." In *The Hollywood Film Industry*. Edited by Paul Kerr, 245–75. London: Routledge, 1986.

al-Hadari, Ahmad. *Tarikh al-sinima fi Misr: al-juz al-awwal min bidayat 1896 ila akher 1930*. Cairo: Nadi al-Sinima bil-Qahira, 1989.

Haggard, Stephan. *Pathways from the Periphery: The Politics of Growth in the Newly Industrializing Countries*. Ithaca, NY: Cornell University Press, 1990.

Hall, Peter A. *Governing the Economy: The Politics of State Intervention in Britain and France*. New York: Oxford University Press, 1986.

Hall, Peter A., and David Suskice, eds. *Varieties of Capitalism: The Institutional Foundations of Comparative Advantage*. New York: Oxford University Press, 2001.

Hansen, Bent, and Girgis A. Marzouk. *Development and Economic Policy in the U.A.R. (Egypt)*. Amsterdam: North-Holland, 1965.

Hansen, Bent, and Karim Nashashibi. *Foreign Trade Regimes and Economic Development: Egypt. Special Conference Series on Foreign Trade Regimes and Economic Development, National Bureau of Economic Research*. Vol. 4. New York: Columbia University Press, 1975.

Harley, John Eugene. *World-Wide Influences of the Cinema: A Study of Official Censorship and the International Cultural Aspects of Motion Pictures*. Los Angeles: University of Southern California Press, 1940.

Harvey, David. *The Condition of Postmodernity: An Enquiry into the Origins of Cultural Change*. Cambridge: Basil Blackwell, 1989.

Hasan, Ilhami. *Muhammed Tal'at Harb: ra'id sina'at al-sinima al-misriyya, 1867–1941*. Cairo: General Egyptian Book Organization, 1986.

Hassanein, Nasser Galal. *al-Ab'ad al-iqtisadiyya li azmat sina'at al-sinima al-misriyya*. Cairo: General Egyptian Book Organization, 1995.

Hatim, Muhammad 'Abd al-Qadir. *al-Siyasa al-thaqafiyya: mabadi' wa dirasat.* Cairo: Higher Specialized Councils, 1984.

Hays, Will H. *The Memoirs of Will H. Hays*. Garden City, NY: Doubleday, 1955.

Hellmuth, William F., Jr. "The Motion Picture Industry." In *The Structure of American Industry: Some Case Studies.* Edited by Walter Adams, 267–304. New York: Macmillan, 1948.

———. "The Motion Picture Industry." In *The Structure of American Industry: Some Case Studies.* Rev. ed. Edited by Walter Adams. New York: Macmillan, 1954.

———. "The Motion Picture Industry." In *The Structure of American Industry: Some Case Studies.* 3rd ed. Edited by Walter A. Adams. New York: Macmillan, 1961.

Helpman, Elhanan, and Paul R. Krugman. *Trade Policy and Market Structure.* Cambridge: Massachusetts Institute of Technology Press, 1989.

Hershfield, Joanne, and David R. Maciel, eds. *Mexican Cinema: A Century of Film and Filmmakers.* Wilmington, DE: Scholarly Resources, 1999.

Heuer, Federico. *La industria cinematográfica mexicana.* Mexico City: privately printed, 1964.

Hoekman, Bernard M., and Michel M. Kostecki. *The Political Economy of the World Trading System: The WTO and Beyond.* 2nd ed. New York: Oxford University Press, 2001.

Hirst, Paul, and Jonathan Zeitlin. "Flexible Specialization: Theory and Evidence in the Analysis of Industrial Change." In *Contemporary Capitalism: The Embeddedness of Institutions.* Edited by J. Rogers Hollingsworth and Robert Boyer, 220–39. Cambridge: Cambridge University Press, 1997.

Hollingsworth, J. Rogers, and Robert Boyer, eds. *Contemporary Capitalism: The Embeddedness of Institutions.* Cambridge: Cambridge University Press, 1997.

Hoskins, Colin, Stuart McFadyen, and Adam Finn. *Global Television and Film: An Introduction to the Economics of the Business.* Oxford: Clarendon, 1997.

Hozic, Aida A. *Hollyworld: Space, Power, and Fantasy in the American Economy.* Ithaca, NY: Cornell University Press, 2001.

Huettig, Mae D. *Economic Control of the Motion Picture Industry: A Study in Industrial Organization.* Philadelphia: University of Pennsylvania Press, 1944.

Irwin, Douglas A. *Against the Tide: An Intellectual History of Free Trade.* Princeton, NJ: Princeton University Press, 1996.

Issawi, Charles. *Egypt: An Economic and Social Analysis.* New York: Oxford University Press, 1947.

———. *Egypt at Mid-Century: An Economic Survey.* New York: Oxford University Press, 1954.

———. *Egypt in Revolution: An Economic Analysis.* New York: Oxford University Press, 1963.

Jameson, Fredric. *The Geopolitical Aesthetic: Cinema and Space in the World System.* Bloomington: Indiana University Press, 1992.

Jameson, Fredric, and Masao Miyoshi, eds. *The Cultures of Globalization*. Durham, NC: Duke University Press, 1998.

Jarvie, Ian. *Hollywood's Overseas Campaign: The North Atlantic Movie Trade, 1920–1950*. Cambridge: Cambridge University Press, 1992.

Johnson, Randal. *The Film Industry in Brazil: Culture and the State*. Pittsburgh, PA: University of Pittsburgh Press, 1987.

Joseph, Gilbert M., Anne Rubenstein, and Eric Zolov, eds. *Fragments of a Golden Age: The Politics of Culture in Mexico Since 1940*. Durham, NC: Duke University Press, 2001.

Kann, Red, ed. *Motion Picture Almanac, 1951–52*. New York: Quigley, 1952.

Katzenstein, Peter J., ed. *Between Power and Plenty: Foreign Economic Policies of Advanced Industrial Countries*. Madison: University of Wisconsin Press, 1978.

Keohane, Robert O., and Helen V. Milner, eds. *Internationalization and Domestic Politics*. Cambridge: Cambridge University Press, 1996.

Kerr, Malcolm. *The Arab Cold War: Gamal Abdel Nasser and His Rivals*. 3rd ed. New York: Oxford University Press, 1971.

Kerr, Paul, ed. *The Hollywood Film Industry*. London: Routledge, 1986.

Kindem, Gorham. *The American Movie Industry: The Business of Motion Pictures*. Carbondale: Southern Illinois University Press, 1982.

Kindleberger, Charles. *The World in Depression, 1929–39*. Berkeley: University of California Press, 1973.

King, John. *Magical Reels: A History of Cinema in Latin America*. 2nd ed. New York: Verso, 2000.

Kitschelt, Herbert, Peter Lange, Gary Marks, and John D. Stephens. *Continuity and Change in Contemporary Capitalism*. Cambridge: Cambridge University Press, 1999.

Klaprat, Cathy. "The Star as Market Strategy: Bette Davis in Another Light." In *The American Film Industry*. Rev. ed. Edited by Tino Balio, 351–76. Madison: University of Wisconsin Press, 1985.

Laird, Sam, and Alexander Yeats. *Quantitative Methods for Trade-Barrier Analysis*. New York: New York University Press, 1990.

Landau, Jacob. *Studies in the Arab Theater and Cinema*. Philadelphia: University of Pennsylvania Press, 1958.

Levy, Clement, ed. *Stock Exchange Yearbook of Egypt*. Cairo: n.p., 1937.

Liebes, Tamar, and Elihu Katz. *The Export of Meaning: Cross-Cultural Readings of Dallas*. New York: Oxford University Press, 1990.

Litman, Barry R. *The Motion Picture Mega-Industry*. London: Allyn and Bacon, 1998.

Lowry, Edward G. "Trade Follows the Film," *Saturday Evening Post* 198 (November 7, 1925): 12.

Mabro, Robert. *The Egyptian Economy, 1952–1972*. Oxford: Clarendon, 1974.

Mabro, Robert, and Samir Radwan. *The Industrialization of Egypt 1939–1973: Policy and Performance*. Oxford: Clarendon, 1976.

Mahfouz, Medhat. "Les salles de projections dans l'industrie cinématographique." In *Egypte: 100 ans de cinema*. Edited by Magda Wassef, 124–29. Paris: Editions Plume and Institute du Monde Arabe, 1995.

Malkmus, Lizbeth, and Roy Armes. *Arab and African Film Making*. London: Zed Press, 1991.

Mara'i, Farida, ed. *Sahafat al-sinima fi Misr: al-nisf al-awwal min al-qarn al-'ishrin*. Ministry of Culture, National Film Center. Cinema Files (1). Cairo: Lotus, 1996.

Mattelart, Armand. *Multinational Corporations and the Control of Culture: The Ideological Apparatuses of Imperialism*. Atlantic Highlands, NJ: Humanities Press, 1979.

Maxfield, Sylvia, and Ben Ross Schneider, eds. *Business and the State in Developing Countries*. Ithaca, NY: Cornell University Press, 1997.

al-Mazzaoui, Farid, ed. *Index du cinéma égyptien, 1952–1953*. Vol. 1. Cairo: Imprimerie Costa Tsoumas, 1953.

———. *Index des films égyptiens, 1953–1955*. Vol. 2. Cairo: Imprimerie Costa Tsoumas, 1955.

———. *Index des films égyptiens, 1955–1962*. Vol. 3. Cairo: Imprimerie Costa Tsoumas, 1962.

Migdal, Joel S., Atul Kohli, and Vivienne Shue, eds. *State Power and Social Forces: Domination and Transformation in the Third World*. Cambridge: Cambridge University Press, 1994.

Miller, Toby. "The Crime of Monsieur Lang: GATT, the Screen, and the New International Division of Cultural Labor." In *Film Policy: International, National, and Regional Perspectives*. Edited by Albert Moran, 72–84. London: Routledge, 1996.

Milner, Helen. *Resisting Protectionism: Global Industries and the Politics of International Trade*. Princeton, NJ: Princeton University Press, 1988.

Mirabile, Lisa, ed. *International Directory of Company Histories*. Vol. 2. Chicago: St. James, 1990.

Moley, Raymond. *The Hays Office*. New York: Bobbs-Merrill, 1945.

Monsiváis, Carlos. "Mexican Cinema: Of Myths and Demystifications." In *Mediating Two Worlds: Cinematic Encounters in the Americas*. Edited by John King, Ana M. López, Manuel Alvarado, 139–46. London: British Film Institute, 1993.

Monaco, James. *How to Read a Film: The World of Movies, Media, and Multimedia: Art, Technology, Language, History, Theory*. 3rd ed. New York: Oxford University Press, 2000.

Monush, Barry, ed. *International Motion Picture Almanac, 1995*. 66th ed. New York: Quigley.

Mooney, Michael Macdonald. *The Ministry of Culture: Connections among Art, Money and Politics*. New York: Simon and Schuster, 1980.

Mora, Carl J. *Mexican Cinema: Reflections of a Society, 1896–2004*. 3rd ed. Jefferson, NC: McFarland, 2005.

———. *Mexican Cinema: Reflections of a Society, 1896–1988*. Rev. ed. Berkeley: University of California Press, 1989.

Moran, Albert. "Terms for a Reader: Film, Hollywood, National Cinema, Cultural Identity and Film Policy." In *Film Policy: International, National, and Regional Perspectives*. Edited by Moran, 1–19. New York: Routledge, 1996.

Mosharrafa, M. M. *Cultural Survey of Modern Egypt*. Pt. 2. London: Longmans, Green, 1948.

Moul, Charles, ed. *A Concise Handbook of Movie Industry Economics*. Cambridge: Cambridge University Press, 2005.

Mumtaz, I'tidal. *Mudhakkirat raqibat al-sinima 30 'amman*. Cairo: General Egyptian Book Organization, 1985.

Mursi, Ahmad Kamel, and Magdi Wahba. *A Dictionary of Cinema/Mu'gam al-fann al-sinima'i*. Cairo: General Egyptian Book Organization, 1973.

Netzer, Dick. *The Subsidized Muse: Public Support for the Arts in the United States*. Cambridge: Cambridge University Press, 1978.

Ninkovich, Frank A. *The Diplomacy of Ideas: U.S. Foreign Policy and Cultural Relations, 1938–1950*. New York: Cambridge University Press, 1981.

Noam, Eli M., and Joel C. Millonzi, eds. *The International Market in Film and Television Programs*. Norwood, NJ: Ablex, 1993.

Nye, Joseph, Jr. *Bound to Lead: The Changing Nature of American Power*. New York: Basic Books, 1990.

O'Brien, Patrick. *The Revolution in Egypt's Economic System: From Private Enterprise to Socialism, 1952–1965*. Oxford: Oxford University Press, 1966.

O'Hagan, John W. *The State and the Arts: An Analysis of Key Economic Policy Issues in Europe and the United States*. Cheltenham, UK: Edward Elgar, 1998.

Ohmae, Kenichi. *The Borderless World: Power and Strategy in the Interlinked Economy*. New York: Harper Business, 1990.

Olson, Mancur. *The Logic of Collective Action: Public Goods and the Theory of Groups*. Cambridge, MA: Harvard University Press, 1965.

Oommen, M. A., and K. V. Joseph. *Economics of Indian Cinema*. New Delhi: Oxford and IBH, 1991.

Paranaguá, Paulo Antonio. *Mexican Cinema*. London: British Film Institute and IMCINE, 1995.

Pérez Turrent, Tomás. "Crises and Renovations." In *Mexican Cinema*. Edited by Paulo Antonio Paranaguá, 94–116. London: British Film Institute and IMCINE, 1995.

Pineda, Alexandra and Antonio Paranaguá, "Mexico and Its Cinema." In *Mexican Cinema*. Edited by Paulo Antonio Paranaguá, 15–62. London: British Film Institute and IMCINE, 1995.

Pines, Jim, and Paul Willemen, eds. *Questions of Third Cinema*. London: British Film Institute, 1989.

Polanyi, Karl. *The Great Transformation*. Boston: Beacon Press, 1957 [1944].

———. "The Economy as Instituted Process." In *Trade and Markets in the Early Empires: Economies in History and Theory*. Edited by Karl Polanyi, Conrad M. Arensberg, and Harry W. Pearson, 239–70. Chicago: Regnery, 1971 [1957].

Powell, Walter W., and Paul J. Dimaggio, eds. *The New Institutionalism in Organizational Analysis*. Chicago: University of Chicago Press, 1991.

Prindle, David F. *Risky Business: The Political Economy of Hollywood*. Boulder, CO: Westview Press, 1993.

Puttnam, David. *The Undeclared War: The Struggle for Control of the World's Film Industry*. New York: HarperCollins, 1997.

el-Qassass, Mohammed. "Theatre and Cinema." In *Cultural Life in the United Arab Republic*. Edited by Mustafa Habib. Cairo: UAR National Commission for UNESCO, 1968.

Quinn, Ruth-Blandina M. *Public Policy and the Arts: A Comparative Study of Great Britain and Ireland*. Aldershot, UK: Ashgate, 1998.

Rachty, Gehan, and Khalil Sabat. *Importation of Films for Cinema and Television in Egypt*. UNESCO, 1979.

de los Reyes, Aurelio. "The Silent Cinema." In *Mexican Cinema*. Edited by Paulo Antonio Paranaguá, 63–78. London: British Film Institute, 1995.

Rogowski, Ronald. *Commerce and Coalitions: How Trade Affects Domestic Political Alignments*. Princeton, NJ: Princeton University Press, 1989.

Rosenberg, Emily S. *Spreading the American Dream: American Economic and Cultural Expansion, 1890–1945*. New York: Farrar, Straus, and Giroux, 1982.

Rudolph, Lloyd I. "Establishing a Niche for Cultural Policy: An Introduction." In *Cultural Policy in India*. Edited by Lloyd I. Rudolph. Delhi: Chanakya Publications, 1984.

Sa'ad, 'Abd al-Mun'im. *al-Sinima al-misriyya fi mawsim. 1967–68 to 1979*. Cairo: al-Ahram al-Tijariyya and General Egyptian Book Organization, 1969–1980.

———. *Mujaz tarikh al-sinima al-misriyya*. Cairo: Matabi' al-Ahram al-Tijariyya, 1976.

el-Sadat, Anwar. *In Search of Identity*. New York: Harper and Row, 1977.

Sadat, Jehan. *A Woman of Egypt*. New York: Simon and Schuster, 1987.

Sadoul, Georges, ed. *The Cinema in the Arab Countries*. Beirut: UNESCO and Interarab Centre of Cinema and Television, 1966.

Sawaf, Elie, ed. *Egyptian Trade Index, 1944*. Cairo: E. R. Schindler, 1945.

Schiller, Herbert I. *Communication and Cultural Domination*. White Plains, NY: International Arts and Sciences Press, 1976.

Schnitman, Jorge. *Film Industries in Latin America: Dependency and Development*. Norwood, NJ: Ablex, 1984.

Scott, Allen. *On Hollywood: The Place, The Industry*. Princeton, NJ: Princeton University Press, 2004.

Sharaf al-Din, Durria. *al-Siyasa wal-sinima fi Misr, 1961–1981*. Cairo: Dar al-Shuruq, 1992.

al-Sharqawi, Galal. *Risala fi tarikh al-sinima al-'arabiyya* [*A Treatise on the History of the Arab Cinema*]. Cairo: al-Misriyya, 1970. Translation of PhD diss., Institut des hautes études cinématographiques, 1962.

Shukri, Mohammed, ed. *al-Sijill al-misri lil-masrah wal-sinima*. Pt. 1. Cairo: Matba'a 'Ataya, 1945.

Sklar, Robert. *Movie-Made America: A Cultural History of American Movies*. New York: Random House, 1975.

Smith, Adam. *The Wealth of Nations*. New York: G. P. Putnam's Sons, 1904 [1776].

Solanas, Fernando, and Octavío Getino. "Towards a Third Cinema." In *Twenty-Five Years of the New Latin American Cinema*. Edited by Michael Chanan, 17–27. London: British Film Institute, 1983.

Spruyt, Hendrik. *The Sovereign State and Its Competitors*. Princeton, NJ: Princeton University Press, 1994.

Squire, Jason E., ed. *The Movie Business Book*. 2nd ed. New York: Simon and Schuster, 1992.

Sreberny-Mohammadi, Annabelle. "The Global and the Local in International Communications." In *Mass Media and Society*. Edited by James Curran and Michael Gurevitch, 118–38. 2nd ed. London: Arnold, 1996.

Stanley, Robert H. *The Celluloid Empire: A History of the American Movie Industry*. New York: Hastings, 1978.

Steinmo, Sven, and Kathleen Thelen. "Historical Institutionalism in Comparative Politics." In *Structuring Politics: Historical Institutionalism in Comparative Analysis*. Edited by Sven Steinmo, Kathleen Thelen, and Frank Longstreth, 1–32. Cambridge: Cambridge University Press, 1992.

Stone, Marla Susan. *The Patron State: Culture and Politics in Fascist Italy*. Princeton, NJ: Princeton University Press, 1998.

Sulayman, Muhammad Hilmy. *al-Mujtama' wal-sinima*. Cairo: Ministry of Culture and National Guidance, 1961.

Tawfiq, Sa'ad al-Din. *Qissat al-sinima fi Misr: dirasa naqdiyya*. Cairo: Dar al-Hilal, 1969.

Thabet, Madkour. *Egyptian Film Industry*. Cairo: Ministry of Culture, 1998.

Thompson, Kristin. *Exporting Entertainment: America in the World Film Market, 1907–34*. London: British Film Institute, 1985.

Thoraval, Yves. *Regards sur le cinéma égyptien*. Beirut: Dar el-Machreq, 1975.

Throsby, David. *Economics and Culture*. Cambridge: Cambridge University Press, 2001.

Tignor, Robert L. *State, Private Enterprise, and Economic Change in Egypt, 1918–1952*. Princeton, NJ: Princeton University Press, 1984.

al-Tilmisani, Kamil. *Safir Amrika bil-alwan al-tabi'iyya*. Cairo: Dar al-Fikr, 1957.

Tirole, Jean. *The Theory of Industrial Organization*. Cambridge: Massachusetts Institute of Technology Press, 1988.

Tomlinson, John. *Cultural Imperialism: A Critical Introduction*. Baltimore: Johns Hopkins University Press, 1991.

Trumpbour, John T. *Selling Hollywood to the World: U.S. and European Struggles for Mastery of the Global Film Industry, 1920–1950*. Cambridge: Cambridge University Press, 2002.

Tunstall, Jeremy. *The Media Are American: Anglo-American Media in the World*. New York: Columbia University Press, 1977.

UNESCO. *World Communications: A 200-Country Survey of Press, Radio, Television, and Film*. Paris: UNESCO, 1975.

Van Hemel, Annemoon, Hans Mommaas, and Cas Smithuijsen, eds. *Trading Culture: GATT, European Cultural Policies and the Transatlantic Market*. Amsterdam: Boekman Foundation, 1996.

Vargas, Fernando Macotela. *La industria cinematográfica mexicana: estudio juridico y economico*. Mexico City: National Autonomous University of Mexico, 1969.

Vasey, Ruth. *The World According to Hollywood, 1918–1939*. Madison: University of Wisconsin Press, 1997.

Vitalis, Robert. "American Ambassador in Technicolor and Cinemascope: Hollywood and Revolution on the Nile." In *Mass Mediations: New Approaches to Popular Culture in the Middle East and Beyond*. Edited by Walter Armbrust, 269–91. Berkeley: University of California Press, 2000.

Vogel, Harold L. *Entertainment Industry Economics: A Guide for Financial Analysis*. 6th ed. Cambridge: Cambridge University Press, 2004.

———. *Entertainment Industry Economics: A Guide for Financial Analysis*. 2nd ed. Cambridge: Cambridge University Press, 1998.

Wachtel, David. *Cultural Policy and Socialist France*. New York: Greenwood, 1987.

Wade, Robert. "Globalization and Its Limits: Reports of the Death of the National Economy Are Greatly Exaggerated." In *National Diversity and Global Capitalism*. Edited by Suzanne Berger and Ronald Dore, 60–88. Ithaca, NY: Cornell University Press, 1996.

Wahba, Magdi. *Cultural Policy in Egypt*. Paris: UNESCO, 1972.

Wallerstein, Immanuel. *The Modern World-System*. New York: Academic Press, 1974.

———. "Culture as the Ideological Battleground of the Modern World-System." In *Global Culture: Nationalism, Globalization, and Modernity*. Edited by Mike Featherstone, 31–55. London: Sage, 1990.

Walt, Stephen M. *The Origins of Alliances*. Ithaca, NY: Cornell University Press, 1987.

Wasko, Janet. *Movies and Money: Financing the American Film Industry*. Norwood, NJ: Ablex, 1982.

Wassef, Magda, ed. *Egypte: 100 ans de cinéma*. Paris: Editions Plume and Institute du Monde Arabe, 1995.

Waterbury, John. *The Egypt of Nasser and Sadat: The Political Economy of Two Regimes*. Princeton, NJ: Princeton University Press, 1983.

Wildman, Steven S., and Stephen E. Siwek. *International Trade in Films and Television Programs*. Washington, DC: American Enterprise Institute, 1988.

Wilkins, Mira. *The Maturing of Multinational Enterprise: American Business Abroad from 1914 to 1970*. Cambridge, MA: Harvard University Press, 1974.

Wilson, Rob, and Wimal Dissanayake, eds. *Global/Local: Cultural Production and the Transnational Imaginary*. Durham, NC: Duke University Press, 1996.

Zukin, Sharon. *Landscapes of Power: From Detroit to Disney World*. Berkeley: University of California Press, 1991.

Periodicals

Aksoy, Asu, and Kevin Robins. "Hollywood for the 21st Century: Global Competition for Critical Mass in Image Markets." *Cambridge Journal of Economics* 16, no. 1 (March 1992): 1–22.

Albin, Michael W. "Official Culture in Egypt and the Role of the Book." *Journal of the American Research Center in Egypt* 24 (1987): 71–79.

American Academy of Political and Social Science. *Annals*. November 1926.

———. *Annals*. November 1947.

Baker, Raymond William. "Egypt in Shadows: Films and the Political Order." *American Behavioral Scientist* 17, no. 3 (January–February 1974): 393–423.

———. "Combative Cultural Politics: Film Art and Political Spaces in Egypt." *Alif* 15 (1995): 6–38.

Bradbury, Malcolm. "What Was Post-Modernism? The Arts in and after the Cold War." *International Affairs* 71, no. 4 (October 1995): 763–74.

Central Bank of Egypt. "Development of Egypt's Foreign Trade During the Period 1952–1978." *Economic Bulletin* 19, nos. 3–4 (1979): 233–55.

Crabbs, Jack, Jr. "Politics, History, and Culture in Nasser's Egypt." *International Journal of Middle East Studies* 6, no. 4 (1975): 386–420.

Epstein, David, and Sharon O'Halloran. "The Partisan Paradox and the U.S. Tariff, 1877–1934." *International Organization* 50, no. 2 (Spring 1996): 301–24.

"Globalization: The Interweaving of Foreign and Domestic Policy-Making." Special issue, *Government and Opposition* 28, no. 1 (Spring 1993): 143–51.

Goff, Patricia M. "Invisible Borders: Economic Liberalization and National Identity." *International Studies Quarterly* 44, no. 4 (December 2000): 533–62.

Gourevitch, Peter. "The Second Image Reversed: The International Sources of Domestic Politics." *International Organization* 32, no. 4 (Autumn 1978): 881–912.

Haas, Peter. "Do Regimes Matter? Epistemic Communities and Mediterranean Pollution Control." *International Organization* 43, no. 3 (Summer 1989): 377–404.

Haqqi, Musa. "Quelques aspects économiques du cinéma égyptien." Special issue, *La Revue internationale du cinéma*, no. 16 (1953): n.p.

———. "Azmat al-sinima al-misriyya." *al-Jumhuriyya*, no. 768 (January 1956): n.p.

Hayward, Susan. "State, Culture, and the Cinema: Jack Lang's Strategies for the French Film Industry, 1981–93." *Screen* 34, no. 4 (Winter 1993): 380–91.

Hatem, Abdul Kader. "Cultural Planning and the Features of Our Cultural Revolution." *Egyptian Political Science Review/al-Majalla al-misriyya lil-'ulum al-siyasiyya* 26 (May 1963): 3–15.

"Hollywood on the Nile." *The Egyptian Economic and Political Review* 1, no. 1 (September 1954): 24–25.

Katzenstein, Peter J., Robert O. Keohane, and Stephen D. Krasner. "*International Organization* and the Study of World Politics." *International Organization* 52, no. 4 (Autumn 1998): 645–85.

Kelada, Naguib. "L'Evolution du tarif douanier égyptien." *L'Egypte contemporaine*, no. 325 (July 1966): 41–109.

Kitchelt, Herbert. "Industrial Governance Structures, Innovation Strategies, and the Case of Japan: Sectoral or Cross-National Comparative Analysis?" *International Organization* 45, no. 4 (Autumn 1991): 453–93.

Krasner, Stephen. "State Power and the Structure of International Trade." *World Politics* 28, no. 3 (April 1976): 317–47.

Lancaster, John. "Egyptians to Relive Movie Past." *Washington Post*, January 10, 1997, p. A23.

Langford, Barry, and Douglas Gomery. "Studio Genealogies: A Hollywood Family Tree." *Gannett Center Journal* 3, no. 3 (Summer 1989): 157–75.

Littoz-Monnet, Annabelle. "European Cultural Policy: A French Creation?" *French Politics* 1, no. 3 (November 2003): 255–78.

———. "The European Politics of Book Pricing." *West European Politics* 28, no. 1 (January 2005): 159–81.

Lowry, Edward G. "Trade Follows the Film." *Saturday Evening Post*, no. 198 (November 7, 1925).

Mansfield, Edward, and Mark Busch. "The Political Economy of Non-Tariff Barriers: A Cross-National Analysis." *International Organization* 49, no. 4 (Autumn 1995): 723–49.

North, C. J. "Our Foreign Trade in Motion Pictures." *Annals of the American Academy of Political and Social Science* 128 (November 1926): 100–108.

Organisation Catholique Internationale du Cinéma. "Le Cinéma égyptien." Special issue, *La Revue internationale du cinéma*, no. 16 (1953): n.p.

Ruggie, John. "International Regimes, Transactions, and Change: Embedded Liberalism in the Postwar Economic Order." *International Organization* 36, no. 2 (Spring 1982): 379–415.

Sa'ad el-Din, Mursi. "Egyptian Cinema." *Arts and the Islamic World* 2, no. 2 (Summer 1984): 62–63, 77–78.

Shafer, D. Michael. "Sectors, States, and Social Forces: Korea and Zambia Confront Economic Restructuring." *Comparative Politics* 22, no. 2 (January 1990): 127–50.

Shepsle, Kenneth. "Studying Institutions: Some Lessons from the Rational Choice Approach." *Journal of Theoretical Politics* 1, no. 2 (April 1989): 131–47.

Storper, Michael. "The Transition to Flexible Specialization in the U.S. Film Industry: External Economies, the Division of Labor, and the Crossing of Industrial Divides." *Cambridge Journal of Economics* 13, no. 2 (1989): 273–305.

———. "Flexible Specialisation in Hollywood: A Response to Aksoy and Robins." *Cambridge Journal of Economics* 17, no. 4 (1993): 479–84.

Storper, Michael, and Susan Christopherson. "Flexible Specialization and Regional Industrial Agglomerations: The Case of the U.S. Motion Picture Industry." *Annals of the Association of American Geographers* 77, no. 1 (1987): 104–17.

Strauss, William Victor. "Foreign Distribution of American Motion Pictures." *Harvard Business Review* 8, no. 3 (April 1930): 307–15.

"The Ten Commandments." *The Egyptian Economic and Political Review* 1, no. 3 (November 1954): 12.

Waterman, David. "The Structural Development of the Motion Picture Industry." *American Economist* 26, no. 1 (1982): 16–27.

Zayida, Gamal. "Min wara' tadahwur sina'at al-sinima fi Misr." *al-Ahram al-iqtisadi*, no. 642 (May 1981).

Government Publications

Egypt, Central Agency for Public Mobalization and Statistics (CAPMAS). *al-Ihsa'at al-thaqafiyya: al-sinima wal-masrah*. 1964–1996. Cairo.

Egypt, Consultative Council. *Nahwa siyasa thaqafiyya lil-insan al-misri*. Report of the Services Committee. Cairo, December 1985.

Egypt. *Customs Tariff*. Tawfik M. Sadek, trans. Cairo: MELES, 1986.

Egypt. *Customs Tariff (Harmonized)*. Tawfik M. Sadek, trans. Cairo: MELES, 1994.

Egypt, General Egyptian Organization for the Cinema, Broadcasting, and Television. *al-Sinima al-misriyya/ Le Cinéma égyptien/ Egyptian Cinema/ El Cine egipcio*. Farid al-Mazzaoui, ed. No. 1. Cairo, 1963.

———. *Bahth: itijahat wa raghbat al-jumhur fil-aflam al-'arabiyya wal-ajnabiyya wa takhtit al-intaj al-sinima'i fil-qita' al-'am*. Technical Office. Cairo, February 1964.

———. *Bahth: aflam al-mawsim al-sinima'i 1963–64 wa iyradatuha, mashru'at al-aflam lil-mawsim al-sinima'i 1964–65 lil-qita'ayn al-'am wal-khas*. Technical Office. Cairo, September 1964.

Egypt, General Egyptian Organization for the Cinema and Broadcast Engineering. *Bahth: aflam al-mawsim al-sinima'i 1964–65 wa iyradatuha, mashru'at al-aflam lil-mawsim al-sinima'i 1965–66*. Technical Office. Cairo, October 1965.

———. *Bahth: dur al-sinima fil-Jumhuriyya al-'Arabiyya al-Mutahida*. Technical Office. Cairo, April 1965.

Egypt, General Egyptian Organization for the Cinema, Broadcasting, and Television. *Sina'at al-sinima: haqa'iq w-arqam*. Technical Office. Cairo, March 1964.

Egypt, General Egyptian Organization for the Cinema and Broadcast Engineering. *Sina'at al-sinima: haqa'iq w-arqam*. Technical Office. Cairo, February 1965.

Egypt, General Egyptian Organization for the Cinema. *Sina'at al-sinima: haqa'iq w-arqam*. Research and Statistics Administration. Cairo, March 1967.

———. *Sina'at al-sinima: haqa'iq w-arqam*. Research Administration. Cairo, April 1971.

Egypt, General Egyptian Organization for the Cinema, Theatre, and Music. *Sina'at al-sinima: haqa'iq w-arqam*. General Inspectorate for Festivals and Cinematic Research. Cairo, May 1975.

———. *Sina'at al-sinima: haqa'iq w-arqam*. Cairo, 1979.

———. *The Motion Picture Industry in Egypt*. Cairo, May 1979.

Egypt, General Egyptian Organization for the Cinema. *Aflam al-mawsim al-sinima'i 1968–1969*. Research Administration. Cairo, September 1969.

———. *Aflam al-mawsim al-sinima'i 1969–1970*. Research Administration. Cairo, October 1970.

Egypt, Ministry of Culture. *al-Sijill al-thaqafi/ Sijill al-thaqafa*. Supreme Council for Culture, General Administration for Planning and Mobilization. Cairo, 1948–1954; 1959–1960; 1972–1977; 1979–1990; 1992; 1994–1996.

———. *4 Mu'tamarat min ajli thaqafa qawmiyya ishtirakiyya insaniyya. No. 1: al-sinima, al-masrah, al-kitab, al-fan al-tashkiliyya*. Cairo: Dar al-Kitab al-'Arabi, 1967.

———. *Nahwa intilaq thaqafi fi funun al-masrah wal-musiqa wal-sinima wal-kitab wal-funun al-jamila*. Cairo: Dar al-Kitab al-'Arabi, 1967.

———. *al-Siyasa al-thaqafiyya: bayan al-duktur Tharwat 'Ukasha wazir al-thaqafa amam lajnat al-khidmat bi majlis al-umma fi 16 yunyu 1969*. Cairo: Dar el-Kutub, 1969.

———. *Cinema in the United Arab Republic*. Cairo, 1970.

———. *Taqrir al-majlis al-qawmi lil-thaqafa wal-funun wal-adab wal-i'lam*. Specialized National Councils, Supreme Council for Culture. Cairo, various years: "al-Sinima fi Misr: waqi'ha wa mustaqbalha." 1st Session, 1979–1980; "Mustaqbal sina'at al-sinima." 3rd Session, 1981–1982; "al-Sinima wa dawruha fil-ta'lim wal-thaqafa." 4th Session, 1982–1983; "Da'am wa tatwir anshitat al-thaqafiyya." 7th Session, Sept. 1985–June 1986; "al-Hurriyya wal-thaqafa fi hayatina." 8th Session, Sept. 1986–June 1987; "Siyasa tatwir sina'at al-sinima al-misriyya." 10th Session, 1988–1989.

———. *Atlas al-khidmat al-thaqafiyya, 1993*. General Administration for Planning and Continuation. Cairo, 1994.

Egypt, Ministry of Education. *al-Nashra al-thaqafiyya al-misriyya*. Vol. 4. General Administration for Culture. Cairo, 1956.

Egypt, Ministry of Finance, Statistical Department. *Annual Statement of the Foreign Trade of Egypt*. 1918 through 1960. Cairo, 1919–1963.

Egypt, Ministry of National Economy, Statistical Department. *Annuaire Statistique, 1945–1946 and 1946–1947*. Cairo, 1951.

———. *Annuaire Statistique, 1947–1948 and 1948–1949*. Cairo, 1952.

Egypt, Specialized National Councils, *Taqrir al-majlis al-qawmi lil-thaqafa wal-funun wal-adab wal-i'lam*. Cairo, 1979–1991.

Egypt, Supreme Council for the Promotion of the Arts, Literature, and Social Sciences. *al-Taqrir al-sanawi, 1958–59*. Cairo.

Nigeria. *Cultural Policy for Nigeria*. Lagos: Federal Government Printer, 1988.

Mexico, Department of Interior. *Ley y reglamento de la industria cinematográfica*. Mexico City, 1962.

Pakistan, National Commission on History and Culture. *The Cultural Policy of Pakistan*. Islamabad: Pagemaker, 1995.

United States Department of Commerce. National Telecommunications and Information Administration. *Globalization of the Mass Media*. Washington, DC: Government Printing Office, 1993.

United States Department of Commerce. *Review of Foreign Film Markets during 1936*. Nathan D. Golden, Bureau of Foreign and Domestic Commerce, Motion Picture Section. Washington, DC: April 1937.

United States, Records of the Department of State Relating to the Internal Affairs of Egypt. National Archives, General Services Administration. Washington, DC. Roll 15, Microcopy 571, Social Matters 883.40, 1910–1929.

United States, Confidential U.S. State Department Central Files (declassified). Egypt: Internal Affairs and Foreign Affairs. University Publications of America, Frederick, Maryland. Record Group 59, Reel 8, Motion Pictures 883.4061, 1945–1949; Record Group 59, Reel 29, Motion Pictures 874.452, 1950–1954;

Record Group 59, Reel 7, Economic Matters 611.74, 1955–1959; Record Group 59, Reel 14, Motion Pictures 874.452, 1955–1959.

Journals

Cine Film/Cine-Orient. 1948–1960.
al-Sinima wal-tarikh 1–3, nos. 1–12, 1992–1994.

Industry Publications and Records

Cinema Industry Chamber of Commerce. *al-Taqrir al-sanawi*. Multiple years.
Federation of Egyptian Industries. *Yearbook*. Cairo, 1960–1985.
———. Annuaire. Cairo, 1955–56; 1958–59.
———. *al-Kitab al-sanawi*. Cairo, 1987–88/1988–89.
———. *al-Kitab al-dhahabi*. Cairo, 1972.
———. *Livre d'or de la Féderation Egyptienne de l'Industrie*. Cairo: Imprimerie Schindler, 1948.
Motion Picture Association of America. "Trade Barriers Report (Egypt)." 1991–1998.
———. "Market Background—Egypt." 1989.
Pascal, Jacques, ed. *The Middle-East Motion Picture Almanac/ al-Dalil al-sinima'i lil-sharq al-awsat*. 1st ed. Cairo: S.O.P. Press, 1947.
———. *Annuaire du cinéma pour le Moyen Orient et l'Afrique du Nord*. Cairo: Imprimerie La Patrie, 1949.
———. *Annuaire du cinéma pour le Moyen Orient et l'Afrique du Nord*. Cairo: Imprimerie La Patrie, 1951–1952.
———. *Annuaire du cinéma pour le Moyen Orient et l'Afrique du Nord*. Cairo: Imprimerie La Patrie, 1954.

Index